Thermal Spray Coating Applications in the Chemical Process Industries

MTI Publication No. 42

Richard P. Krepski

Intermet Technology

Georgetown, PA

Published for the Materials Technology Institute of the Chemical Process Industries, Inc. by NACE International

Library of Congress Catalog Card Number: 93-86263
ISBN: 1-877914-59-2
Printed in the United States of America

Published for The Materials Technology Institute of the Chemical Process Industries, Inc. by

NACE International
P.O. Box 218340
Houston, TX 77218-8340

Information on availibility of this publication can be obtained from NACE International

About MTI

The Materials Technology Institute of the Chemical Process Industries, Inc. (MTI) is a unique, cooperative research and development organization representing private industry. Its objective is to conduct generic, nonproprietary studies of a practical nature on the selection, design, fabrication, testing, inspection, and performance of materials and equipment used in the process industries.

Through membership in MTI, companies can solve nonproprietary problems of major concern to the process industries, leverage research and development dollars by participating in the direction and results of MTI studies, capitalize on the expertise of member company representatives, and learn about MTI's accomplishments in a timely fashion.

More information about membership in MTI can be obtained from the Executive Director. MTI members receive copies of MTI publications as part of their membership.

The Materials Technology Institute of the Chemical Process Industries, Inc.

1215 Fern Ridge Parkway, Suite 116
St. Louis, MO 63141-4401
Tel: 314/576-7712
Fax: 314/576-6078

Contents

List of Tables

List of Figures

Acknowledgments

The thorough editorial input of Dr. Chris Berndt of State University of New York at Stony Brook (SUNY) is gratefully acknowledged. The comments of R.W. (Bud) Ross, former MTI technical consultant, and the MTI Resource Group for Project No. 60 are also appreciated. ASM International kindly arranged release of figures from various publications as well as reprinting of the 1993 "Thermal Spray Buyers' Guide." Industrial contacts providing useful input included Larry Moscowitz of Amoco, Dick Drossman of Wear Management Services, Robert Tucker of Union Carbide, Val Zanchuck and Bob Unger of Hobart-Tafa, Doug Harris of APS-Materials, John Nerz of Metco Perkin-Elmer, Robert Sulit of Duralcan, and Casey Reynolds of National Coatings.

James Q. Lackey was instrumental in initiating and providing direction for this project.

Executive Summary

Thermal spray coating is a versatile technology with the potential for broad utilization in the chemical process industries. Materials properties such as corrosion performance, high-temperature environmental stability, wear resistance, and thermal response can be significantly upgraded by selection of the appropriate coating system and application method.

Effective implementation of thermal spray technology requires an appreciation of the complex relationships between materials selection, coating process variables, resultant microstructure, and performance. This information is presented in a manner for easy use by the nonspecialist. The available application methods and coating materials are reviewed. Industrial experience (case histories) and pertinent research are highlighted. The last section outlines emerging innovations with special relevance to the chemical process industries.

CHAPTER 1

Introduction

Materials performance in many industrial applications is strongly affected by the behavior of surfaces. For example, corrosion response, reactivity with environment (oxidation, sulfidation, dissolution by solvents, etc.), wear characteristics, and thermal properties are all critically dependent on surface character. Aspects of mechanical behavior such as fatigue response and environmentally assisted cracking are also dependent on surface character. Utilization of coatings to upgrade component performance is often a cost-effective alternative to substitution of superior bulk materials. Coating is also a means of refurbishing worn or otherwise degraded parts.

Thermal spraying is perhaps the most versatile of the coatings technologies. In this line-of-sight process, a powder, wire, or rod feed is rapidly brought to molten or near-molten state by a heat source such as oxygen-fuel combustion, electric arc, or electrically induced plasma. The atomized material is then impelled by a gas stream to impact a suitably prepared substrate. The broad spectrum of coating materials (generally termed "feedstocks") includes polymers, metals and alloys, refractory oxides, and ultrahard carbides. Composite coatings can be readily developed by spraying different layers, by spraying blended powders, or by use of feedstocks that are themselves composite in nature. Suitable substrates range from wood and plastic to the high-temperature components of jet turbines and rocket motors.

Thermal spraying is not restricted to in-shop applications; in some cases it can be utilized in the field for the refurbishment of standing structures. Thermal spray coatings may be combined with other coatings to yield a synergistic upgrading of performance.

In addition to coatings applications, thermal spraying is a means of rebuilding worn parts and manufacturing nearly net-shape components, including composites. Driven primarily by aerospace applications, thermal spray technology has advanced rapidly in terms of materials, hardware, and automated process control. The complex relationships between process variables, coating structure, and performance have become more clearly understood, so that optimal coatings can be obtained repeatedly.

The chemical process industries (CPI) provide numerous opportunities for utilization of thermal spray coatings. Potential benefits include greater systems reliability, reduced maintenance costs and downtime, substitution of less expensive bulk materials, and extending the limits of materials performance so that new processes may be instituted.

There have been some examples of successful utilization of thermal spray coatings by these industries. Most notable are the refurbishment of Yankee drier rolls and black liquor recovery boilers in pulp and paper manufacturing and the utilization of hard cemented carbide and ceramic coatings in various pump and valve components. However, it is surprising at first glance that thermal spraying does not have a greater presence in the CPI, considering its extensive use by other demanding industries such as aerospace. For example, the modern jet engine has hundreds of components protected by this method.

The modest utilization of thermal spray coating by the CPI is explained, in part, by the prevalence of applications demanding complete isolation of the substrate from a corrosive environment. The build-up of impinging droplets to form the coating generally gives rise to porosity, networks of oxides, and other defects that detract from the coating's ability to form a continuous, long-term barrier. Corrosion problems are also aggravated by process streams with variable chemistry. Quite often, minor impurities can have profound influence on corrosion response. These factors undoubtedly caused disappointing results for many of the initial plant trials with thermal spray coatings.

Early work on industrial applications was possibly also confounded by the perception on the part of both specifiers and applicators that coating characteristics were defined by the selection of coating composition and method of application. Aerospace and other industries have learned that coating reliability is linked to the control of critical process variables. Funding for industrial materials development is less likely to permit this level of sophistication in evaluation of new options.

Thermal spray coating technology has progressed significantly in recent years. This is indicated dramatically by the escalating attendance at annual conferences and exhibitions sponsored by the Thermal Spray Division of ASM International.(1) There have been major advances in coating materials and methods of application, and also a greater appreciation of the importance of process and coating quality control. There has also been increased focus by applicators on the market potential for coatings applied on-site. Based on these considerations, it is most appropriate for plant engineers to take a closer look at thermal spray, even if past experience has been negative.

This publication is intended to assist in such efforts by providing an overview of process and materials options, an understanding of the factors affecting coating performance, and a synopsis of available information on case histories. Because the field is evolving so rapidly, the concluding section outlines possible future trends of particular relevance to the CPI.

(1) ASM International, Materials Park, OH.

CHAPTER 2

Generic Description of Thermal Spray Coatings

The "Materials Science" Approach

Advances in the ability to analyze structure down to the near-atomic scale have greatly increased our understanding of the behavior of materials in recent years. Evolving from physical metallurgy, the relatively new discipline of "materials science and engineering" has contributed to many of the key advances of modern technology. While classical metallurgy took a largely trial-and-error approach to improvement in properties, materials science focused on the structure of the material and the mechanistic link between structure and performance. Structure is determined by the interplay between composition and processing. All materials—metals, ceramics, plastics, composites (combinations of different materials), and natural materials like stone and wood—can be understood in terms of this basic equation:

composition + processing structure = properties

These properties, in turn, translate to performance parameters such as useful life, failure mode, and environmental stability in an engineered component.

The maturation of thermal spray coating as a technology has, to a large extent, been caused by adoption of the materials science approach. The modern jet turbine is the prime example of thermal spraying being effectively utilized. Hundreds of engine components requiring predictable behavior under extreme service conditions have been reliably coated, and the adoption of new coatings continues to extend the envelope of engine performance. This success represents the efforts of numerous materials science professionals employed by both users and suppliers. Research results from universities and government laboratories have also been rapidly translated into practice. Of course, the levels of funding associated with defense- and aerospace-related endeavors has a great deal to do with this success. Applications in the other great "driver" of new technology, the automotive industry, have been less extensive, but there are good examples of thermal spray applications in the internal combustion engine.

With some notable exceptions, industrial utilization of thermal spray coatings has lagged behind that in the higher-technical sectors. Plant personnel are generally not well-versed in the principles of materials science. Funds may not be available to perform the pilot scale studies necessary to transfer an innovation from the lab to commercial service. There is sometimes a poor rapport between corporate research and production units. Vendors have made great claims about the performance of coatings in service environments that are complex and not completely understood, leading to failures and reduced user-enthusiasm for exploration of further options.

If the potential benefits of thermal spray are to be more extensively realized by the CPI, plant technical personnel need to have an appreciation of the materials science issues involved in coating performance. Thus, a basic understanding of the key aspects of coating structure is essential.

The Process of Coating Development

Thermal spray coatings are applied by a variety of systems:

- conventional combustion ("flame-spray");

• electric arc spray;

• plasma (including chamber spraying with controlled atmosphere);

• high-velocity oxy-fuel (HVOF); and

• detonation (D-Gun†);

These are considered in detail in Section 4. Regardless of the delivery hardware and form of feedstock (powder, wire, or rod), thermal spraying involves projection of molten or semimolten particles against the surface of the material to be coated (the "substrate"). The impinging particles may spread to form pancake-like "splats" or "lamellae," or partially fragment into smaller droplets. Figure 2.1 diagrams the dynamics of the forming splat and indicates a complex interplay of heat transfer and material flow as the particle's motion is suddenly arrested.

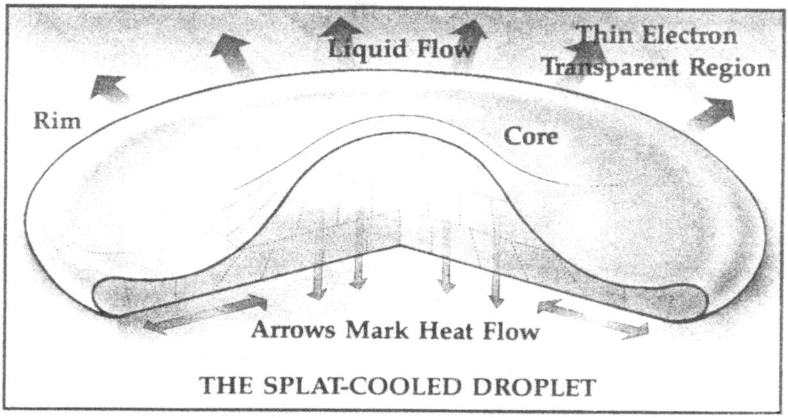

Figure 2.1 Model of a "Splat" as it Impacts Substrate. Reproduction courtesy APS-Materials; original artwork from SUNY at Stony Brook.

Typical lamellae dimensions are 1 to 5 μm thick and 50 to 100 μm diameter.[1,2] The buildup of splats results in a highly oriented structure, with lamellae parallel to the substrate surface. Another

† Trade name.

consideration is the formation of shock waves in both the particle and substrate[3]—the physics of which is beyond the scope of this book. The integrity of the coating will be dictated by the interactions of the splats with the substrate (adherence or adhesive strength) and with each other (coherence or cohesive strength).

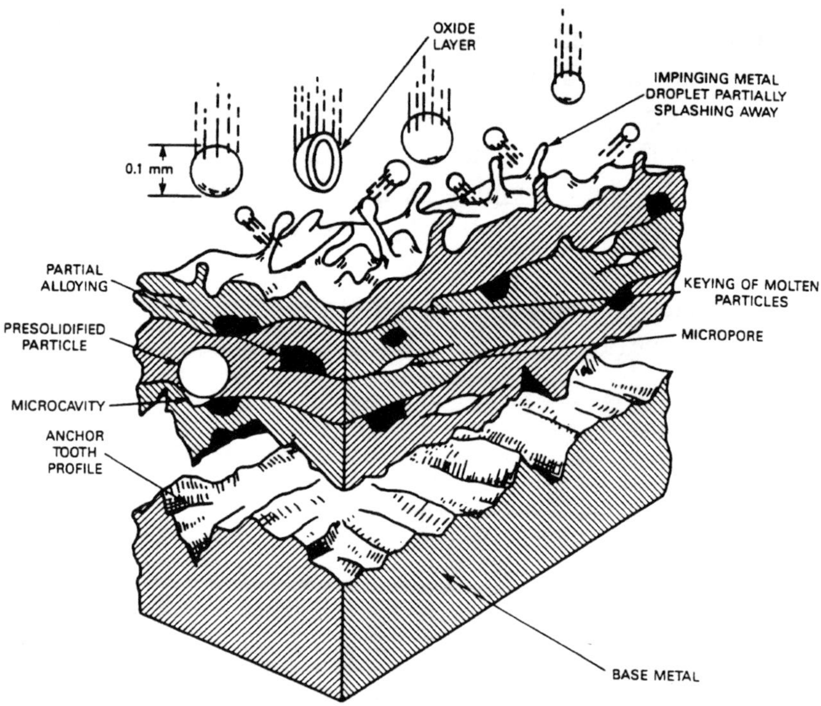

Figure 2.2 Generic coating structure;[4] used with permission of ASM International (original artwork from SUNY at Stony Brook).

Figure 2.2 depicts the typical features of the generic thermal spray coating. The impinging particle (if not an oxide to begin with) is likely to interact with the surrounding atmosphere during its time of flight, and entrapped oxide is often the demarcation of the boundary between splats. This boundary is also associated with significant porosity. In fact, recent work indicates that perhaps only 20 to 30 percent of the interface between splats in a typical coating can be viewed as a "microweld."[1,5] The cohesion of the coating is, to some extent, derived from the interlocking fingers of the individual splats. Some of the impinging particles may not be molten on impact, and this results in coating inhomogeneity and additional porosity. The microstructure of the splat interior may itself be nonuniform. Sometimes this is intended, as in the case of wear-resistant coatings of carbides held in a tough metallic matrix (Figure 2.3). However, the brief residence time of the feed in the heating zone of the spray device may also result in undesirable coating inhomogeneities that reflect nonuniformity of the feed itself. Figure 2.2 also suggests the importance of the surface profile of the substrate in determining bond strength.

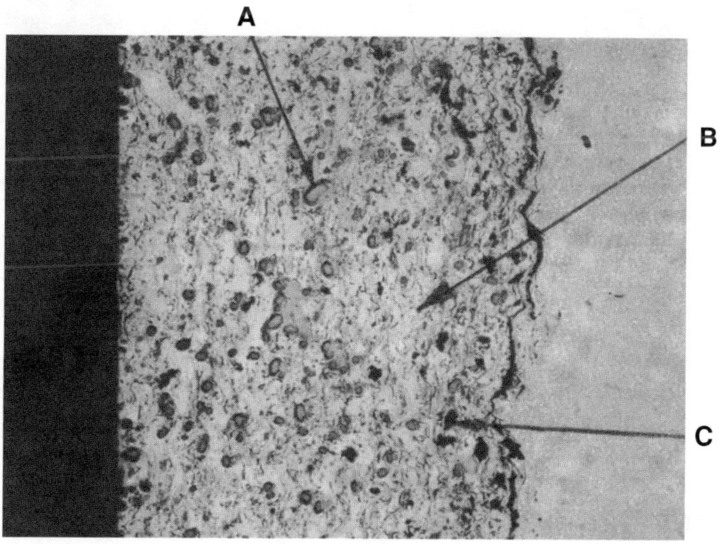

Figure 2.3 Scanning electron photomicrograph of plasma-sprayed aluminum bronze/titanium diboride on copper (right) after sectioning and polishing. The boride particles (A) are randomly distributed in the bronze matrix (B). The polished surface also reveals porosity (C).[6] Used with permission of ASM International.

Figure 2.4 is a coating cross section illustrating the orientations of the individual splats. In this case, the feedstock was a blend of two different powders—an iron (Fe)-based alloy resistant to adhesive wear and a nickel (Ni)-based powder added to promote adhesive and cohesive strength. The chemical differences between the two alloys give rise to significant contrast when viewed in the scanning electron microscope, such that the individual splats can be distinguished.

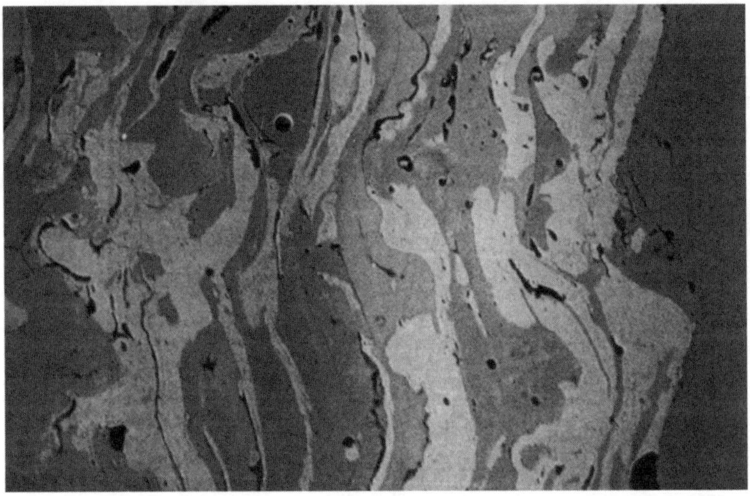

Figure 2.4 Microstructure of Tribolite† "One-Step" coating on steel. The high contrast between adjacent splats was caused by the use of a blend of a wear-resistant, iron-based powder and a bond-enhancing, nonferrous powder in the feedstock.[7] Courtesy of Wear Management Services.

Bonding of Coating to Substrate

Establishment of the proper microroughness of the substrate surface, usually accomplished by abrasive blasting, is critical for obtaining adequate adhesion of the coating. The irregular surface anchors the initial layer of spray particles and also redistributes the residual and service stresses at the coating/substrate interface so the coating is less likely to spall. This mechanical interlocking is one of the mechanisms for adherence, but other physical and chemical bonding forces also come into play. In rare cases, a true metallurgical bond is established, akin to that encountered in hot dip coating and liquid metal joining methods.[8] In most cases, the brevity of the time at elevated temperature suggests that weaker atomic attractions of the kind responsible for normal wetting are in operation.

When considering behavior at the coating/substrate interface, one should keep in mind that even a freshly blast-cleaned metal surface will generally have an oxide film that is invisible to the naked eye. This is important in the case of stainless steels and oxidation-resistant, high-temperature alloys which rely on a self-healing, protective oxide film (usually chromium oxide) to impart environmental stability. However, carbon steel surfaces also have this tendency to form oxide when in contact with the atmosphere. In soldering, oxide film is eliminated through the use of fluxes. In thermal spraying, this is not an option. Surface preparation for spray coating eliminates thicker oxide scales and other contaminants that would interfere with bonding. Organic contaminants may be more effectively removed with solvents. Blasting residues may, in turn, be a source of coating contamination.

The optimum substrate surface profile will be a function of specific coating material and the delivery system. Clearly, size distribution and spreading tendency of the impinging particles is a consideration. Selection of substrate profile is also a function of targeted coating thickness and desired final coating surface profile.

Certain metallic feedstocks, termed "bond coats" or "self-bonding," are capable of bonding effectively to clean substrates with smooth surface profiles. This is useful in applications where the optimum substrate surface profile cannot be established or where superior bonding is critical to performance. In other cases, the combination of coating material and spraying system may yield unacceptably poor adherence unless an intermediate bond layer is first established.

Currently, the most commonly employed bond coats are formulations based on nickel-aluminum (Al). Other corrosion resistant bond coats include nickel-chromium (Cr), aluminum-bronze, and molybdenum (Mo). The "M-CrAlY" formulations, with M corresponding to iron, nickel and/or cobalt (Co), are commonly utilized as bond coats for the high-temperature oxide thermal barrier coatings used in jet turbine applications. In this application, accommodation of thermal stresses and high-temperature oxidation resistance are key considerations.

Historically, the advantageous bonding characteristics of nickel-aluminum coatings have been ascribed to heat generated by an exothermic reaction between nickel and aluminum,[8] and the term "exothermic coating" is often utilized. However, recent work calls into question the importance of this reaction.[5,9] Bond coats are discussed in detail in Section 7.

Coating Defects

The principal coating defects influencing performance are indicated in Figure 2.2. For applications involving corrosion, coating porosity is a primary consideration, since a continuous path to the substrate will lead to premature failure unless the coating is galvanically sacrificial. Optimization of process parameters and innovations in spray systems for corrosion applications thus target elimination of through-porosity. Attainment of this objective has been elusive, although great strides have recently been claimed. Progress in this area is a key to more effective utilization of thermal spray in the CPI.

Sealing the coating is a means of mitigating the effects of porosity. Most sealers are simply highly thinned paints. More advanced approaches may involve inducement of chemical reactions within the pores or special treatments by laser or hot isostatic pressing. Certain coating formulations, generally nickel-, cobalt, or iron-based with appropriate additives, are termed "self-fluxing" and are amenable to consolidation by melting. When combustion is used as the heat source for both spraying and melting, this is termed the "spray-and-fuse" approach. In the case of plasma, an arc transferred from gun to substrate can be utilized as the heat source. Spray-and-fuse and plasma transferred arc introduce a significant thermal cycle to the substrate that might influence its structure and properties.

As mentioned above, most metals sprayed in air will react with oxygen during spraying. Droplets in flight can be envisioned as being surrounded by balloon-like skins of oxide. This gives rise to a continuous network of oxide in the coating, which may be detrimental in terms of corrosion, cohesion, and other performance parameters. Oxygen may also react with particle interiors. This may impart enhanced wear response in some cases (molybdenum or aluminum) but may induce embrittlement in others (tantalum [Ta] or titanium [Ti]). Other important reactions with the spray atmosphere include carbon loss for carbide wear coatings and interactions with nitrogen and hydrogen in some cases.

Many of the key recent advances in thermal spray hardware involve use of a controlled atmosphere to reduce porosity, entrapped oxide, and other gas-coating reactions. Chamber spraying may add appreciable cost, and spraying under vacuum (generally termed "low-pressure" spraying) can lead to excessive heating of the substrate.

Surface of the Coating

The texture and roughness of the coating surface are important considerations with respect to wear characteristics. Surface features are influenced by feed particle size (for powder), coating process parameters, and method of application. Roughness of the substrate is also a factor in determining coating surface profile. Many applications demand a surface finished by machining, grinding, or lapping. This can add significantly to the cost. Mechanical finishing may also help seal pores to some extent.

Surface porosity can also provide a means of locking in lubricants, release agents, and other additives; and can enhance the bonding of additional coating layers.

Stress Considerations

Deposition of molten or semimolten particles on a relatively cool substrate will tend to generate residual tensile stresses in the coating as it is quenched. This can result in spalling (detachment at the coating/substrate interface) as the coating rapidly cools, especially if the coating is relatively thick. Besides limiting coating thickness, residual stress influences coating mechanical behavior, wear performance, and corrosion response. Residual stress is a function of delivery system and process parameters. A finer particle size is associated with higher stress levels.[10] One reported advantage of the high-velocity D-Gun[†] system is that it yields coatings under residual compressive stress, at least in the case of tungsten carbide/cobalt wear coatings.[11] Modest preheating of the substrate may reduce residual coating tensile stresses to some extent.

Differences in coating and substrate thermal expansion coefficients are important, particularly in applications where there is a significant thermal cycle. Differential thermal expansion may be accommodated by a bond coat and, in some cases, by a gradual transition from bond coat to exterior coating.

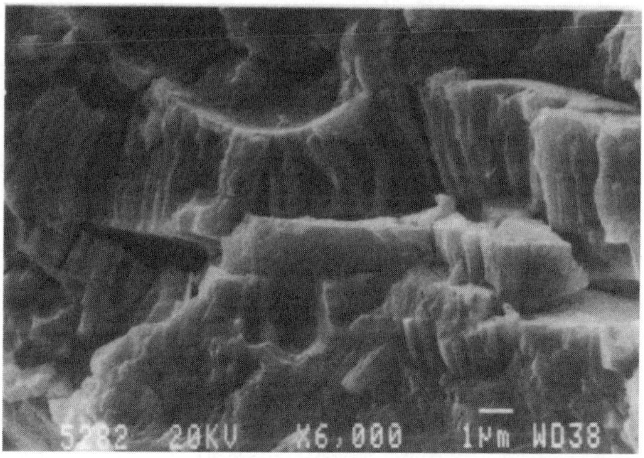

Figure 2.5 Scanning electron micrograph of a zirconia-yttria plasma-sprayed coating after fracture. Note the columnar morphology within each splat, aligned normal to the heat flow direction during the quench. Unbonded areas between splats can also be seen.[13] Used with permission of ASM International.

Stress considerations are especially critical for ceramic coatings, which generally tend to crack under tensile loads. Damage due to impact encountered in normal handling may be a key factor limiting industrial use of ceramic coatings. The generally low thermal conductivity of ceramics can promote cracking due to differential thermal stresses—this is termed "thermal shock." As in advanced structural ceramic components, ceramic coatings may be toughened by a range of microstructural manipulations, including controlled microcracking to blunt crack propagation, fiber additions, islands of tougher phases, and partially amorphous (glassy) structures. In "reaction toughening" of structural ceramics like "partially stabilized zirconia," tensile stresses at the tip of a crack trigger a phase change that puts the crack tip in compression and opposes further propagation. However, behavior of rapidly quenched coatings may be quite different from that of the same material in bulk form.[12] The fracture surface of a zirconia coating is shown in Figure 2.5.

Microstructural Considerations

Review of the methods of toughening ceramic coatings brings into focus a key issue from the materials science perspective—the influence of processing on the microstructure of the coating material. Each splat can be looked at as a microcasting that undergoes a rapid quench. For metallic thermal spray coatings, cooling rates on the order of one million degrees Celsius per second (10^6 °C/s) can be expected. Such rapid solidification can give rise to microstructures (and properties) markedly different from those of bulk materials produced by casting, wrought fabrication, or consolidation of powders. Stable phase structures may not have the opportunity to form. Solutes that normally separate into different phases may remain dissolved in the matrix. For some metallic alloy compositions, solidification may occur so quickly that the nucleation and growth of crystals to generate the normal grain structure of the metal do not have the opportunity to occur. Instead, a glassy or amorphous structure analogous to that of a silicate glass may be obtained. This structure can have properties vastly different from its crystalline counterpart. These various "metastable" structures may convert to more stable forms during service.

Figure 2.6 illustrates the influence of rapid quenching on the structure and properties for a powder combustion-sprayed zinc-18% nickel coating on steel. The portion of the splat that was most rapidly quenched when it impacted the surface has a uniform appearance and shows severe cracking. Areas that solidified more slowly display the segregations more typical of a cast alloy and appear to be much less prone to cracking.

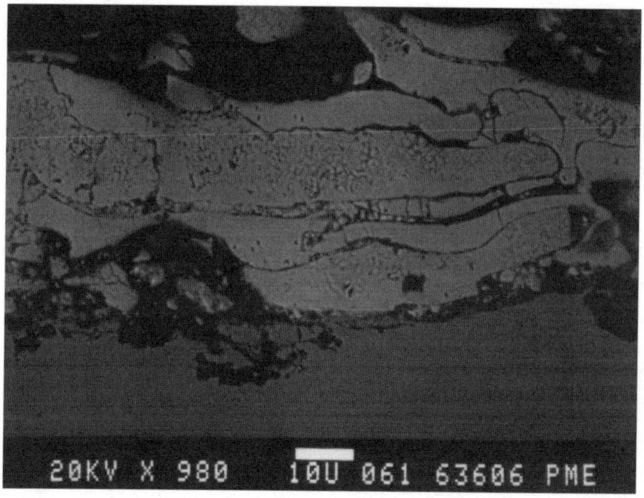

Figure 2.6 Combustion-sprayed zinc-18% nickel on steel. Regions experiencing a faster quench do not show segregations and are more prone to cracking. Courtesy of Intermet Technology.

The partially amorphous character of an as-sprayed, HVOF (high-velocity oxy-fuel combustion) tungsten carbide/cobalt coating is shown in Figure 2.7 (following page). Heat treatment of the coating transforms the amorphous matrix to more stable cobalt-bearing carbides, which upgrades the abrasive wear response.

For ceramics, the effect of quench rate is complicated by the relatively poor thermal conductivities of these materials compared to metals. Splats adjacent to the metal substrate quench more rapidly than those nearer the coating exterior. The outer shell of each splat solidifies more rapidly than the interior. These effects may give rise to inhomogeneous microstructures. For instance, thin glassy films (as depicted in Figure 2.8) have been observed at splat exteriors for sprayed zirconia[15] and at the coating/substrate interface for pure alumina.[16] Microstructures of individual crystals in a metastable glassy matrix similar to that in Figure 2.7 may be encountered for some oxide alloys.

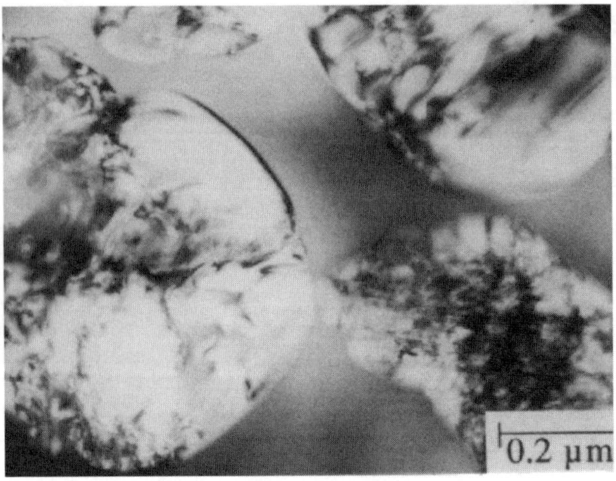

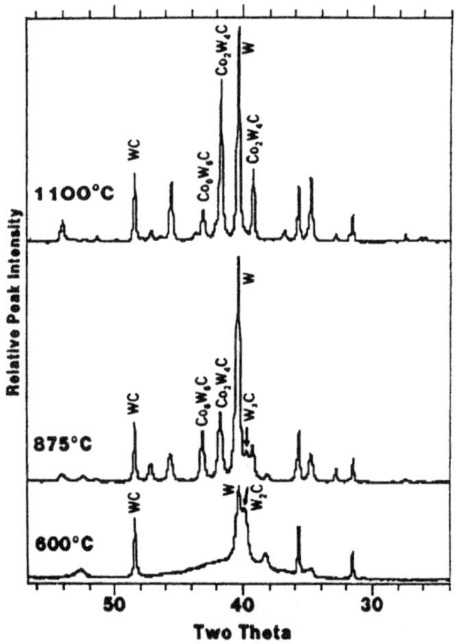

Figure 2.7 Amorphous tungsten carbide/cobalt HVOF coating. Transmission electron micrograph (top) shows amorphous character of matrix of as-sprayed HVOF tungsten carbide/cobalt, with tungsten carbide islands present. X-ray diffraction (bottom) reveals transformation of the amorphous material to more stable cobalt-bearing phases by heat-treatment.[14] Used with permission; prints courtesy Metco Perkin-Elmer.

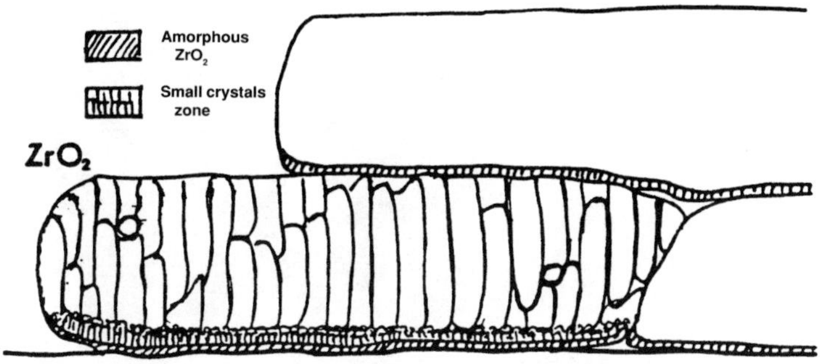

Figure 2.8 Schematic of zirconium oxide splats on a NiCrAlY bond coat. Rapid local quenching leads to an amorphous skin on each splat.[15] Used with permission of ASM International.

Because of the complex interplay between microstructure and quench rate, the properties of a sprayed coating may not match those of the conventionally processed bulk material. For instance, a sprayed austenitic stainless steel may show ferromagnetic behavior, indicating the presence of a ferritic component. The optimum composition for toughening and thermal fatigue resistance of sprayed "partially stabilized zirconia" may not match that for the sintered product.

Another source of coating inhomogeneity on a microscopic scale is the nonuniformity of the starting material. Many thermal spray feeds are simple powder blends or composite powders with an exterior coating on each particle. A recent innovation in wire feeds is the wire sheath with a core of hard, wear-resistant particulate. Prealloyed powders generated from the liquid may also have segregations that evolve during solidification of the powder. In the short time at-temperature in thermal spraying, homogenization of the particle cannot be expected, even if the particle is completely molten. Figure 2.9 shows the microscopic segregation in a partially melted nickel-based particle caught

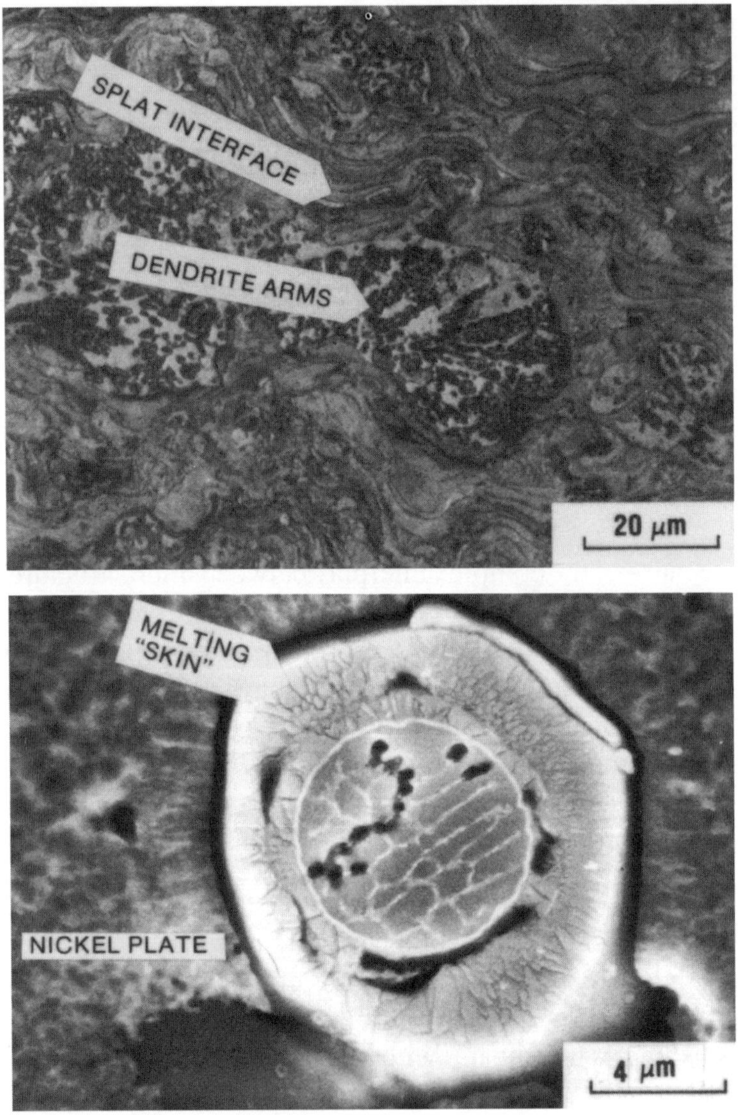

Figure 2.9 Melting behavior of nickel-based superalloy René 80[†] during plasma-spraying and its influence on coating structure. Molten skin and segregated core of particle caught in flight (top) and resultant segregation in the deposited coating (bottom).[17] Used with permission.

during flight and the corresponding inhomogeneity in the coating. This is one reason why the exact nature of the feed material—not just composition—can have a profound influence on properties of the coating.

CHAPTER 3

Key Process Variables

Recent reviews of thermal spraying technology allude to an evolution "from art to science."[18,19] This maturation has been marked by a better understanding of the relationships between process variables and coating quality. For the plasma-spray technique, over a hundred interacting parameters are recognized as having an influence on the coating.[18,20] Some of the major factors are outlined by Kubel in Figure 3.1.[21] Optimization and control of these parameters are the keys to attaining reliable, high-quality coatings.

It is useful to focus on coating structure as the link between process and performance, since this is how the effects of coating parameter variation are manifested.

Chemical Uniformity of the Feedstock

As previously discussed, definition of a target coating composition is not the only consideration involved in selection of the feedstock. For processes utilizing powder, feedstock of a given composition may be commercially available in a variety of forms. Prealloyed powders may be manufactured by gas or water atomization of liquid phase, or by crushing brittle cast ceramics, carbides, or brittle metal alloys. The feed for thermal spraying will reflect the microsegregations that may evolve during solidification. This inhomogeneity may be reflected in the final coating since short time at temperature for the particle in flight does not yield complete homogenization. Another category of powder

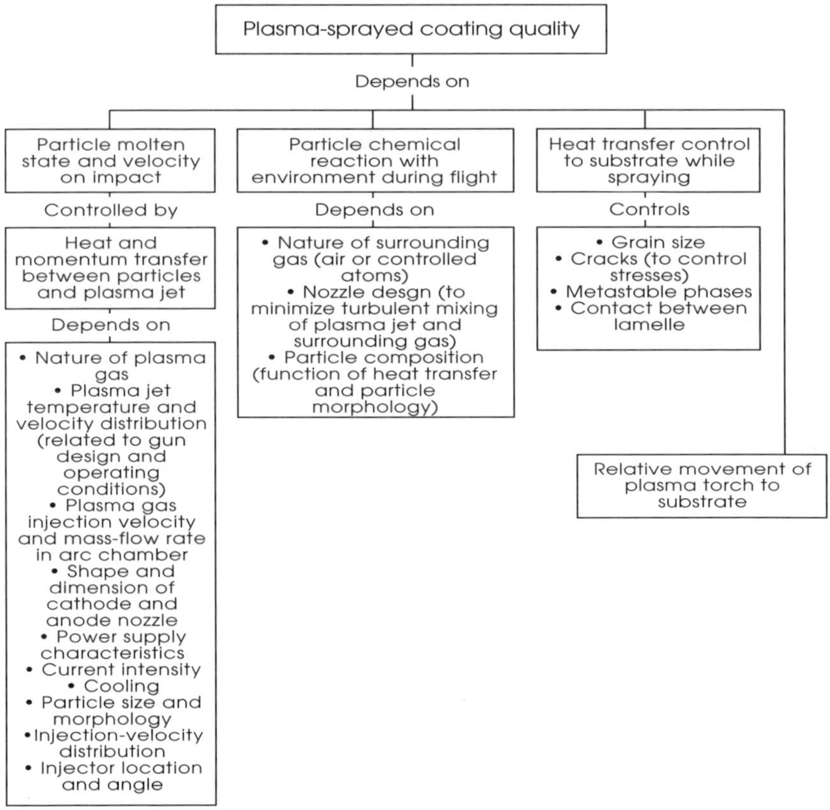

Figure 3.1 Factors influencing plasma-sprayed coating quality. Used with permission of ASM International.

feedstock is agglomerated (spray-dried) products in which fine powders are held together using an organic binder. The agglomerated product may also be sintered to yield a more coherent particle with partial or complete alloying. Composite powders may be generated by chemical cladding of a core particle with a different composition on the exterior, by coating the core with fine particles using a binder, or by sintering. Again, the thermal spraying process may not homogenize the particulate before it impacts the substrate surface. Some powder feeds are simple blends of different powders. As Figure 2.4 indicates, this will give rise to a difference in composition between adjacent splats.

Wire feedstock materials are generally homogeneous prealloys. However, the use of cored wires containing hard materials such as carbides has been explored, and coatings produced from such feedstock will display segregation of the core component.

Segregation due to inhomogeneity of the spray consumable is not necessarily undesirable. For instance, the aluminum surface of a composite nickel-aluminum bond coat powder may promote wetting of steel surfaces and spreading of the impinging particles. Also, the exothermic reaction between nickel and aluminum during time of flight is thought by some to be a key factor in promoting bonding to the substrate by local superheating.[22] However, recent research suggests that this may be less important than previously thought.[5,9] Another example of desirable chemical inhomogeneity is the family of carbide "cermet" wear coatings. The hard carbide provides wear resistance, while a matrix of cobalt, nickel, or nickel-chrome provides toughness. Coated powder may also provide an advantage by minimizing reactions with the atmosphere during time of flight or by improving heat transfer to the powder from the spray heat source.

Particle Shape

In processes that use a powder feedstock, the shape of the particulate can be an important influence on such processes as feed rate into the spray gun, heat transfer, and atmospheric interaction during time of flight. Gas-atomized powders tend to be spherical or droplet shaped, while water-atomized and crushed powders are angular. Powders based on agglomeration tend to be more porous. For an application like thermal barrier coating (TBC), this may in fact be desired. Figure 3.2 represents the broad

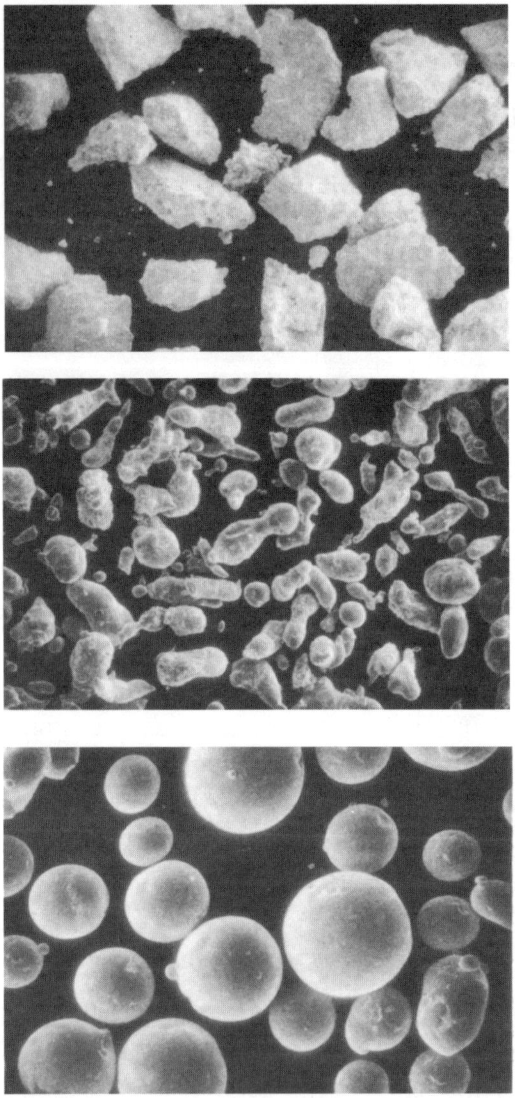

Figure 3.2 Range of thermal spray powder particle shapes: sintered and crushed (top), water-atomized (middle), and gas-atomized (bottom). Courtesy of Metco Perkin-Elmer.

range of powder morphologies that are obtainable. Some novel powder morphologies are becoming available. For instance, a hollow, spherical zirconia "HOSP" powder with enhanced melting characteristics is available for TBC applications.[23]

Spray Particle Size Distribution

The interactions between the spray particle, heat source, atmosphere, and substrate (droplet spreading, quench rate) depend on the size of a given particle. For powder-fed processes, size distribution of the feedstock is essentially the same as that of the droplets impinging the substrate. A tight particle size range is generally desirable, such that spraying parameters can be adjusted to give optimized particle heating. However, the economics of powder production dictate that a broader range of sizes be sprayed. This will increase the proportion of unmelted particles and overheated, excessively oxidized particles. High-speed processes like HVOF require a finer particle size because of the short residence time in the combustion zone. In some cases, a finer particle size will tend to evolve higher residual stresses and lower coating thickness limits,[20] possibly because a faster quench reduces the time during which the coating can relax thermal stresses by plastic deformation. Finer powder may yield a coating that can be ground-finished to a smoother profile. Powder morphology and size distribution are also important with respect to flow from the powder hopper, through the powder delivery system, to the spray torch. This can have a major impact on coating quality and reproducibility.

For wire-fed processes, particle size during time of flight is dependent on such atomizing parameters as atomizing gas flow rate, wire size and feed rate, energy input, and gun configuration. Literature describing the Rokide† rod-feed ceramic combustion

† Trade name.

spray system indicates that larger particles are sprayed than in powder-fed systems and suggests that this leads to greater coating integrity due to impinging particles with higher kinetic energy and greater thermal mass.[24]

Particle Heating

A critical step in *thermal* spray processing is the transfer of energy from the gun energy source to the coating material. Generally, a large fraction of the spray particles should be molten at impact.[25] However, excessive heating above the melting point (termed "superheat") can lead to an undesirable fragmentation of the particles as they strike the surface. Too much superheat can also increase residual stresses,[26] exacerbate oxidation and other atmospheric interactions, and promote selective vaporization that can shift alloy composition. Recent literature on high-velocity oxy-fuel combustion (HVOF) systems indicates that actual particle melting may not be necessary to produce a successful coating.[26-28] The kinetic energy of the particle reportedly acts to "plasticize" the particle on impact[26] and may be converted to heat, inducing local welding to the substrate or to previously deposited splats.[28]

Particle heating from the gun may be subsidized by heat evolved from exothermic reactions within the particle. In particular, the composite nickel-aluminum bond coat feeds are associated with this effect. However, the importance of the exothermic nickel-aluminum reaction has been called into question by recent work.[5,9] Oxidation of aluminum may play a more dominant role.

Particle Velocity

With other parameters optimized, higher particle velocity tends to promote a denser, more adherent coating. This is the essential appeal of the D-Gun and other detonation systems and the basis for recent hardware innovations such as HVOF. Conventional powder combustion gives by far the lowest particle velocity, and this is generally manifested in poor coating properties (low bond strength, high porosity) compared to other powder fed systems.

Mechanistically, the importance of velocity is related to the conversion of particle kinetic energy to other forms of energy on impact and to the spreading out of the droplets into splats. As mentioned above, HVOF and other high-velocity processes are thought to yield better coatings if particles are not molten during flight.[26] Optimized D-Gun coatings may exhibit desirable compressive residual stresses, presumably caused by propagation of compressive shock waves during quenching. The energy of impact may also help break up oxide films on the substrate and particle surface which would otherwise inhibit bonding.

Atmosphere in Flight

Oxidation of particles in flight is generally detrimental to coating integrity for metals, particularly if an oxide skin forms on the droplet surface. Fragmentation and entrapment of these oxides in the coating degrade adhesion and cohesion and give rise to greater coating porosity and intersplat regions more prone to accelerated attack by the environment. Some reactive metals, such as tantalum and titanium, are subject to severe deterioration of mechanical properties because of oxygen pick-up. In many of the carbide-containing wear coatings, the hard carbide phase tends to decompose into other, less desirable phases because of reaction with the atmosphere. Arc-spraying of high-carbon steels for wear reclamation can result in carbon loss as high as 40 percent.[29]

Concern over gas-particle interactions has prompted innovation in thermal spray delivery hardware in recent years, including chamber spraying under reduced pressure or in inert gas,[30,31] employment of various shrouding schemes,[32] and use of inert gas rather than compressed air to accomplish atomization in the arc-spray system.[33]

In some instances, reaction with the atmosphere during spraying is desirable. For instance, the oxygen pick-up during combustion-wire spraying of molybdenum improves wear performance compared to that of plasma-sprayed coatings with lower oxygen content.[34] Oxidation of aluminum is one of the sources of the exothermic heating associated with nickel-aluminum bond coats.[9] Attempts have been made to develop novel coatings by reaction of the droplets in flight. For example, titanium nitride coatings have been generated by plasma-spraying titanium in a nitrogen atmosphere at greater than normal atmospheric pressure.[35]

Substrate Conditions

A properly prepared substrate surface is the key to effective bonding of the coating. The profile introduced by grit blasting provides a mechanical interlock, increases the surface area for chemical and physical bonding, and distributes interface thermal and residual stresses to prevent spalling of the coating. Blasting residues, surface contaminants (such as oils and greases), rust, and gross scale all interfere with bonding.

In general, a metallic surface will re-establish an oxide film promptly after grit blasting or other mechanical treatments. For carbon-steel surfaces, the acceptable interval between blasting and spraying will depend on ambient temperature and humidity. Generally, spraying should be performed within one hour of surface preparation. Excessive preheating in air may also be detrimental.

Preheating does provide a means of reducing residual stresses if it does not degrade the properties of the substrate or severely oxidize the surface. Plasma-spraying in an evacuated chamber (generally termed "low-pressure plasma-spraying") offers the advantage of preheating and cleaning the surface with the plasma torch without concern about substrate oxidation.[2]

Other Independent Variables

Angle of spray impingement is an important consideration. Spraying normal to the substrate is preferred. Too great a departure from this condition will increase coating porosity significantly and decrease deposit efficiency. Figure 3.3 compares the coating structures for powder combustion-sprayed zinc

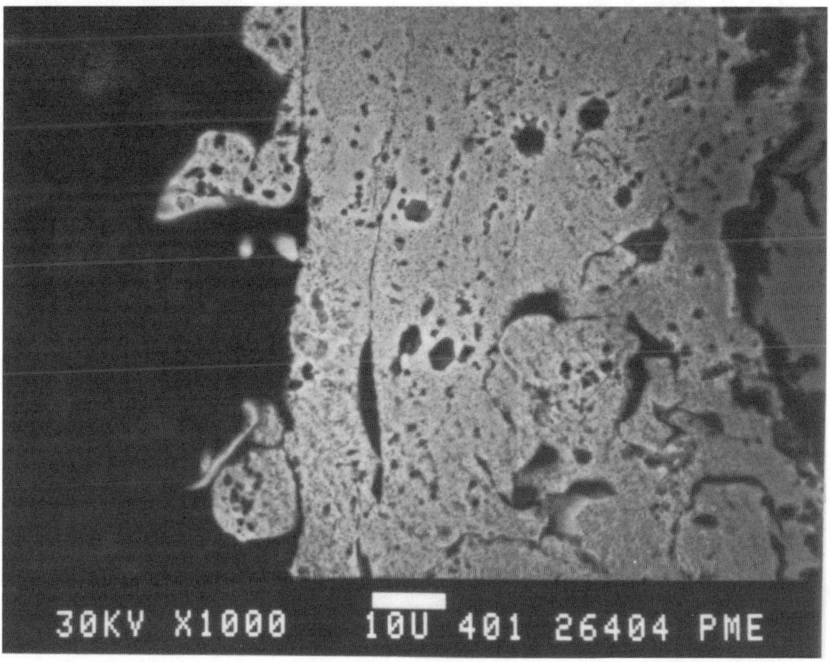

Figure 3.3a Dependence of coating structure on angle of incidence. Zinc powder combustion coating on a flat steel plate. Variable angle of incidence yields appreciable increase in porosity. Courtesy of Intermet Technology.

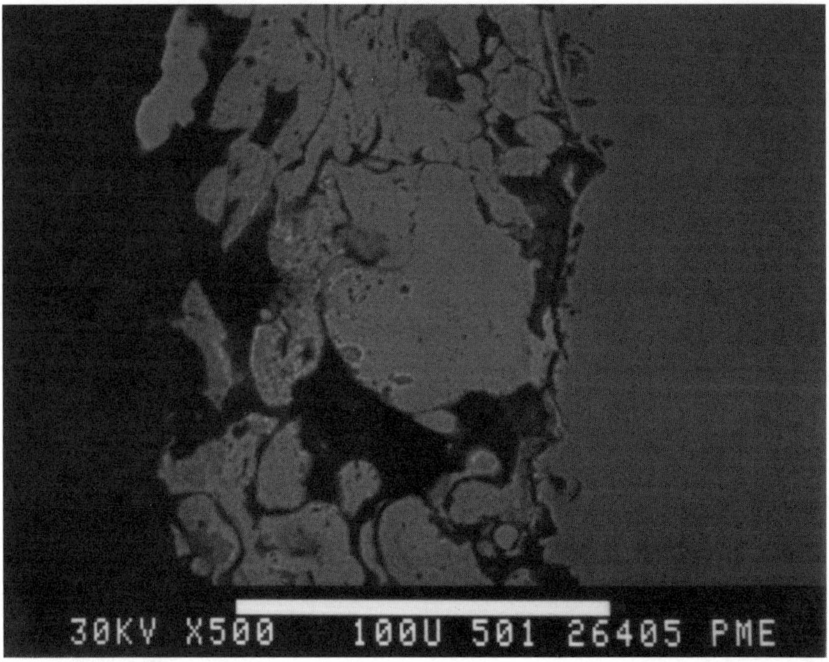

Figure 3.3b Dependence of coating structure on angle of incidence. Comparable coating on ½-in. (1.3-cm) diameter rod rotated in the spray stream. Variable angle of incidence yields appreciable increase in porosity. Courtesy of Intermet Technology.

deposited on a flat steel plate and a ½-in. (1.3-cm) diameter rotating steel bar. Variable angle of incidence in the latter case results in very high porosity. Processes more forgiving with respect to angle of incidence will be more appropriate for coating complex shapes like heat exchangers and for the outer and inner surfaces of pipes. Automated control of gun position via robotics also helps to alleviate problems.

Increasing coating thickness is a means of reducing the likelihood of through-porosity to the substrate and extending coating life when wear and/or uniform corrosion are involved. However, thicker coatings give rise to greater residual stresses, possibly causing poor adhesion or fracture within the coating. Coating buildup needs to be properly controlled by selection of overlap pattern, deposition rate, and relative motion of gun and substrate so that localized overheating and stress problems don't occur. Substrate temperature increase during spraying may be restricted by forced convection using air or carbon dioxide jets.[36]

Control of the stand-off distance between gun and substrate is vital to realization of quality coatings. Factors such as in-flight oxidation, loss of kinetic energy during flight, and particle temperature at impact are related to stand-off distance. Automated torch and substrate positioning will help to avoid problems.

CHAPTER 4

Available Coating Systems

Introduction

Systems for application of thermal spray coatings derive their thermal energy from either gas combustion or conversion of electrical energy. Combustion provides the heating in conventional oxy-fuel combustion (also termed "flame-spraying") and in the recently developed "high-velocity oxy-fuel" (HVOF) systems. The detonation gun (D-Gun)† is also a combustion system in which discrete explosions are used to propel the coating material as well as heat it. Electrical energy is employed to melt the wire feed in arc-spray systems and generate a high-temperature ionized gas in plasma systems. The various delivery strategies are shown schematically in Figures 4.1 through 4.3.

Conventional Combustion Spraying

Acetylene is the most commonly used fuel gas for flame-spraying because of the high-temperature of the oxy-acetylene flame (3,100°C [5,600°F]). Other fuel gases include propane, hydrogen, natural gas, and methylacetylene-propadiene (MAP).[38]

Powder-feed combustion guns are simple, compact, and lightweight. In powder flame-spraying, the flame is used to both melt the feed and direct it to the substrate. Particle velocities are therefore low, and the coating is generally inferior to those obtained with other systems in terms of porosity and bond strength.[38] The simple powder combustion systems are appropri-

† Trade name.

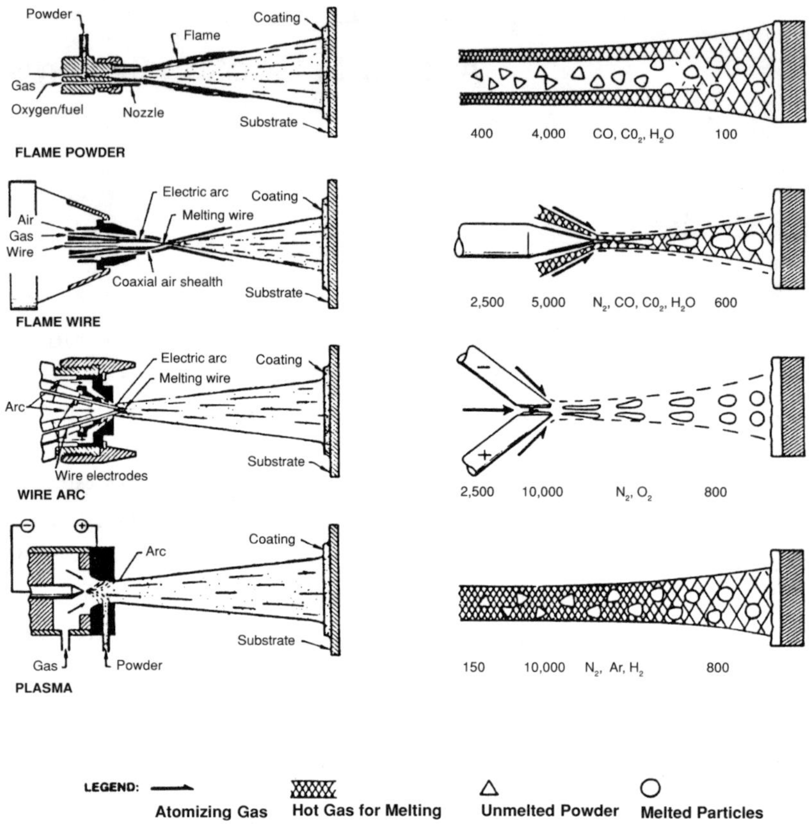

Figure 4.1 Overview of thermal spray processes.[8] Used with permission of ASM International.

ate for the most forgiving maintenance shop applications. They are also extensively used for "spray-and-fuse" applications in which the deposit is densified and metallurgically bonded to the substrate by melting. This is essentially a brazing process. The spray torch may in fact double as the heat source utilized to remelt the coating.

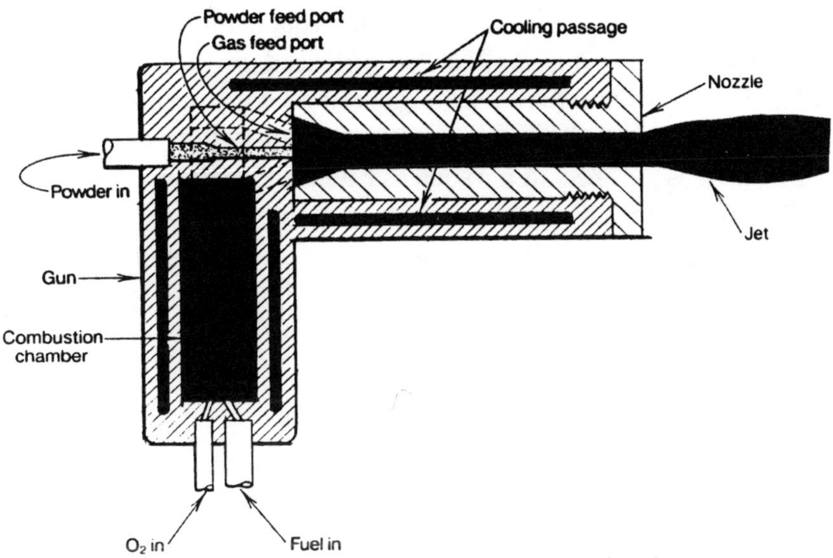

Figure 4.2 Schematic of the Jet Kote† HVOF system.[37] Used with permission of ASM International.

Figure 4.3 Schematic of Detonation Gun (D-Gun).[36] Used with permission.

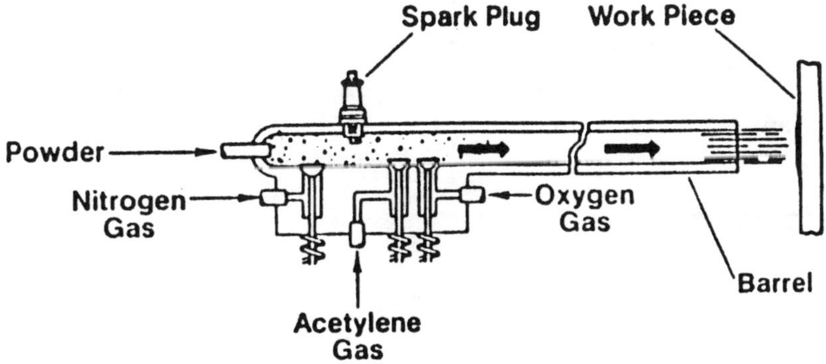

The basic powder-combustion approach may be upgraded significantly by using auxiliary compressed air to accelerate the melted particles and to reduce temperature at the periphery of the flame. The latter is important when spraying vaporization-prone feeds such as zinc. Systems such as Eutectic Terrodyne† 2000 and 3000, and Metco† 5P-II, greatly extend the capabilities of powder combustion processes in terms of coating quality, range of feeds, and deposition rate. There are also spray torches specifically tailored for the low heat input requirements of plastic spraying. An example is the UTPlast F-311† gun marketed by UTP Welding Materials of Houston.

Wire-and rod-feed combustion systems differ from the powder fed systems in that the combustion gas is only used to melt the coating material. Atomization and acceleration to the substrate are accomplished with compressed air. A typical particle velocity is 700 f/s (210 m/s), compared to perhaps 100 f/s (30 m/s) for simple powder combustion. Wire combustion is the system of choice for molybdenum coating because oxygen pick-up during spraying upgrades wear performance.[39] The Norton Rokide† system is specifically designed for combustion spraying of brittle ceramic rods.[24] An air motor or turbine is used to drive the feed. The supplier claims that the system offers a coating quality advantage compared to powder combustion spraying of ceramics without the capital outlay of plasma systems. Another option in combustion spray feedstock is the "cord" feed, in which a continuous length of hollow plastic tubing is filled with powder. This variant is practiced mainly in Europe.[38]

Arc-Spraying

In the standard arc-spray process, two consumable metal wire electrodes are advanced to meet at a point in an atomizing gas stream. The potential difference between the nearly contacting wires generates an arc, which melts the wire tips. A nozzle directs the atomizing gas (usually compressed air) across the arc zone, forming the liquid metal droplets and propelling them to the substrate.[39] The appealing aspects of the arc-wire process include high energy efficiency and deposition rate.[40] Bernecke indicates a typical deposition rate of 30 lb/h (13.5 kg/h), compared to 4 to 20 lb/h (1.8 to 9 kg/h) for powder combustion and 12 lb/h (5.4 kg/h) for wire combustion.[41] Some commercial arc-spray systems can reportedly spray zinc at up to 115 lb/h (52.8 kg/h).[42] The only factors controlling spray rate are the current rating of the system and the wire feed rate necessary to use this current.[39] The absence of a combustion flame also leads to reduced substrate heat-up, which is desirable for some components. Vendor literature for arc-spray systems claims generally superior adhesion and cohesive strength compared to combustion and equates this with greater superheat associated with the arc melting process.[39]

Plasma-Spraying

In the late 1950s, the rapidly evolving aerospace industry demanded a new generation of coatings for high-temperature service involving erosion/corrosion control, thermal barriers, and sealing between moving parts.[39] The coating compositions desired in many applications were refractory oxides and carbides not sprayable by standard low-temperature combustion. Plasma coating systems were developed to meet this need.

In conventional plasma-spraying, a DC arc between the tungsten alloy cathode and copper nozzle anode is utilized to heat a flowing inert gas stream, usually argon. The gas is partially ionized—hence the term "plasma," which refers to the fourth and least-recognized form of matter. Part of the energy stored in the plasma flame is released when the gas "de-ionizes." Also, diatomic gases such as N_2 and H_2 are often added to the plasma gas to increase its enthalpy. These molecular bonds are broken when the gas is excited by the arc, and recombination of the atoms provides further energy for heating the feed material for the coating. In fact, some metal feeds, such as Ni-Cr, provide a catalytic surface for this recombination.[2]

The dynamics of the plasma jet are exceedingly complex and beyond the scope of this work. Influential factors include current density, cold plasma gas flow rate, arc-chamber and nozzle design, and gas thermal conductivity. The plasma jet also draws in surrounding air rapidly by turbulent mixing,[21] but this can be reduced by effective nozzle design. A fundamental issue is how the powder feed is injected into the plasma jet. The powder is delivered using an inert carrier gas and is typically introduced directly into the anode nozzle downstream from the arc root.[21] Injection may be perpendicular or inclined at an angle opposing the flow of the plasma jet. The momentum provided by the powder carrier gas must be sufficient to enable the powder to penetrate into the plasma and be carried at the proper trajectory to the substrate. Particle size distribution is important because it affects injection velocity, momentum transfer, heat transfer, and heat needed to melt and superheat the particle. Even with a very tight particle size distribution, the injection velocity can have a broad range.[21] A significant recent innovation is a torch design providing for axial introduction of feedstock, which reduces variability in the spraying process.[43]

Gas temperature at the core of the plasma is on the order of 17,000°C (30,000°F).[8] The combination of high temperature, fast particle heating, and inert atmosphere of the plasma results in desirable coating characteristics for a broad range of spray materials.[39] Standard equipment has a fairly high feedstock throughput capability (3 kg/h [7 lb/h]), and special high-power guns can attain 25 kg/h (55 lb/h) for certain materials.[17] Bernecki cites particle velocities on the order of 300 m/s (980 ft/s) for a 40 kW system and 500 m/s (1,600 ft/s) for an 80 kW system.[41] The plasma coating technique is the workhorse of the thermal spray industry, particularly in the area of aerospace applications.

In the plasma transferred arc process, an arc is established between the torch electrode and the conductive workpiece, and the plasma-sprayed feedstock develops a molten puddle on the surface, leading to a dense cast microstructure and metallurgical bonding to the substrate.[38] This is essentially a weld overlay technique, with relatively low base-metal dilution (typically on the order of 10 percent). Plasma transferred arc is particularly suitable for hardfacing with cobalt-based alloy powders.[44]

Detonation Gun

The D-Gun detonation gun system was developed in the 1950s by Union Carbide, actually before the initial work on plasma-spray systems.[11] It remains the premium coating technology for advanced wear coating applications and the "target" that innovative systems seek to supersede. Proprietary control of the technology was retained by Union Carbide through a network of Coatings Service shops which are now administered by Praxair Surface Technologies of Indianapolis, Indiana, although some competitive systems have emerged, including the relatively compact FARE[†] gun[45] and recently developed Russian, Japanese, and Swedish systems.[41]

The detonation gun resembles a small cannon.[40] Oxy-acetylene gas and powder are injected into the barrel and ignited with a spark plug. The combustion chamber is then flushed with nitrogen, and a new cycle begins.[2] Operation frequency is typically four to eight cycles per second, giving a relatively low spray rate of 0.5 to 2 lb/h (0.3 to 0.9 kg/h).[41] The gun generates noise on the order of 150 dB.[2] This level is capable of causing "full body damage," so acoustic isolation in a spray chamber is mandatory.

The key to D-Gun operation is spray particle velocity, estimated at 2,500 ft/s (750 m/s). Union Carbide's next generation Super D-Gun† reportedly increases gun pressure by 50 percent, essentially doubling the kinetic energy of the particles.[11] Spray particles emerging from the D-Gun do not need to be completely molten. Shock waves and conversion of kinetic energy to heat on impact generate a well-bonded coating with minimal porosity (less than two percent). Residual stresses can be compressive for tungsten carbide/cobalt coatings, an important advantage for wear and fatigue resistance.[11] Ground and lapped coatings have exquisite surface finish (on the order of one microinch [0.025 µm]), again critical for wear performance. Bernecki has observed that WC-9% Co ("LW1") applied by D-Gun displays the best wear resistance of all thermal spray coatings below 500°C (930°F).[41] Besides the various metal-bonded carbide cermets, refractory oxides such as aluminum oxide ("LA2") may be advantageously sprayed.[41]

HVOF Systems

If the D-Gun is analogous to a cannon, HVOF can be likened to a rocket engine. Browning Engineering developed the original HVOF system, Jet Kote,† to provide an alternative to the proprietary D-Gun technology. The original focus was on carbide cermet coatings for wear resistance in gas turbine applications. In the Jet Kote technology, a fuel gas (propylene, hydrogen, propane, MAP) is injected under pressure with oxygen into a water-cooled combustion chamber. The combustion gases are discharged through four ports in the combustion head which increase gas velocity to hypersonic level and enable the gas to make a 90-degree bend into a water-cooled nozzle. Powder is injected via carrier gas (nitrogen, argon, helium) into the intersection of the four exhaust gas streams, where the powder is evenly distributed into the core of the flame and propelled through the nozzle. The gases exit the nozzle at a hypersonic velocity, as indicated by the presence of visible shock diamonds in the flame. Gas velocities have been measured in the range from 1,500 to 2,000 m/s (4,900 to 6,500 ft/s), on the order of five times the speed of sound.[46] The technical literature indicates particle velocities in the range of 550 to 800 m/s (1,800 to 2,600 ft/s).[47] Flame temperature is relatively low, on the order of 2,900°C (5,250°F),[48] making it difficult to spray ceramics and refractory metals. Since dwell time in the flame is short, heat transfer to large powder particles may not be sufficient. It requires finer powder particle size and tighter particle size distribution than other processes.

The Jet Kote technology has been acquired and refined by Stellite Coatings. The success of Jet Kote has prompted development of other HVOF systems, including Diamond Jet† (Metco), Top Gun† (Miller Thermal), and J-Gun† (Whitco, now acquired by Hobart-Tafa Technologies). The systems differ in terms of nozzle design, combustion fuel, and coating feed injection.

The Top Gun system is capable of utilizing acetylene as the fuel gas, yielding the highest flame temperature of any HVOF gun and permitting spraying of ceramics such as chromium oxide and refractory metals like molybdenum. The J-Gun technology permits use of a liquid fuel (kerosene), and product literature indicates that this option provides the highest combustion pressures and particle velocities of any HVOF system. Wire feed is also an option for the J-Gun system.

As in the case of the D-Gun, high particle impact velocity implies that particles may not need to be molten to attain a high-quality coating. Kreye et al. discuss a deformation-induced surface melting of impinging particles analogous to that encountered in explosive cladding.[49] Metco workers refer to spray particles "plasticized without superheat" to minimize residual stress problems, which is especially important with regard to internal diameters and thicker coatings.[26]

Extensive evaluation of Jet Kote-applied WC-12% Co in a joint study by the United States Air Force and General Electric has indicated coating quality equivalent to that of the D-Gun process.[50] Although developed as an alternative to the D-Gun, HVOF is emerging as a competitor for plasma-spraying in some applications. Besides coating quality and reduced residual stress, advantages include superior deposit efficiency, reduced sensitivity to spray angle, and fewer critical process variables.[50] Gun-to-substrate distance is more flexible and generally larger than that for plasma. A typical range is 5 to 10 in. (12 to 25 cm), compared to 3 to 5 in. (7 to 13 cm) for plasma.[50] A spray "beam" 24 in. (61 cm) long is alluded to in J-Gun product literature. Low flame temperature may result in greater stability of feed composition. For instance, vaporization of the plastic component of aluminum-polyester "clearance control" coatings is far less than for plasma.[50]

Besides the various carbide cermet powders, coating materials successfully applied by HVOF have included alloy 718 (UNS N07718), "Hastelloy† C" (a designation used for both alloy C [UNS N10002] and alloy C-276 [N10276]), Tribaloy† T-400 and T-800 (Stoody Deloro Stellite), copper-nickel-indium, aluminum-polyester, and various other iron-, nickel- and cobalt-based feeds.[50] As mentioned above, the higher flame temperature of the Top Gun reportedly permits use of molybdenum and refractory oxides such as chromium oxide and aluminum oxide. Very interesting results have been obtained for Jet Kote spraying of mixed alumina-40% titania.[48] While plasma-spraying gives rise to complete fusion and formation of aluminum titanate, HVOF results in a composite of alumina embedded in a soft titania matrix. Enhanced wear performance is noted in the latter case.

A compilation of available thermal spray hardware is provided in Appendix A.

CHAPTER 5

Recent Advances in Spray System Technology

A key aspect of the maturation of thermal spray technology has been the refinement of systems for applying coatings. Besides upgrading coating quality, these advances have extended the range of coating chemistries and feedstock forms. Developments have also improved repeatability and reliability of coatings as well as process economics. Some of the most recent include:

Table 5.1

Recent Advances in Spray System Technology

Basic Technology	Innovative Technology	Key Improvement
Plasma	Chamber Spray	
	- Inert Gas	Minimum Oxidation
	- Low Pressure	Minimum Porosity
	Inductively Coupled	Uniform Melting
	Water-Stabilized Plasma	High Spray Rate
	Underwater Spraying	Rapid Quench
Arc-Spray	Inert Gas Atomization	Reduced Porosity
	Chamber Spray	Reactive Metals
	Single Wire	Deposit. Efficiency
	Cored Wire Feed	Hard Wear Coatings
HVOF	Shrouding	Minimum Porosity
	Browning HVAF	Economics
	Browning HVIF	Dense Coatings
D-Gun	Super D-Gun	Residual Compressive Stress/Smooth Finish

In addition to modifications in spray hardware and control of environment, refinements in automation andprocess control have helped to ensure more reliable coatings.

Chamber Spraying

Controlling the ambient atmosphere during coating by spraying in a chamber is a means of improving coating quality. Inert gas spraying prevents oxidation of molten metal particles in flight, enhancing intersplat bonding and eliminating a source of undesirable oxide inclusions. However, the coating may still resemble an air-sprayed deposit in terms of porosity and residual stress.[31] A more radical alteration in process and product is effected by spraying in an evacuated chamber, and this has been the subject of extensive research, particularly in the context of plasma-spraying.[30, 31, 51, 52] Generally termed "low pressure plasma-spraying," this process results in a longer spray jet and higher particle velocity than spraying in air.[31] The substrate can be preheated without oxidation to reduce residual stress in the coating. Heating of the substrate/coating interface may also promote a true chemical bond by diffusion and formation of intermediate phases. The longer flame and lack of convective cooling also promotes annealing effects in the deposited coating, so nonequilibrium phases and amorphous character may not be retained. This may or may not be a disadvantage, depending on its impact on specific properties. The absence of oxide stringers and porosity as a result of entrapped gas leads to significantly improved coating quality. Chamber spraying may also be adapted to provide sputter-cleaning of the substrate prior to coating, significantly upgrading coating adhesion in some cases.[18]

Arc-spraying under low pressure is a challenge because the arc is significantly extended and may cause a short circuit between the gun hardware and the substrate. A special nozzle design overcomes this problem.[30, 53] Low pressure arc-spraying of reactive metals like tantalum and titanium can be successfully accomplished. These are useful coating compositions for some chemical process applications.

Spraying in an inert atmosphere or under reduced pressure is restrictive in terms of capital and operating costs, low production rate, and size of components. Herman has suggested the intriguing possibility of spraying the interiors of large pipes or vessels by using the vessel itself as the controlled atmosphere chamber.[18] This has significant potential for CPI. Advances in robotic systems for remote spraying of complex surface profiles make such an approach a realistic possibility, but it is unclear if this has been accomplished in a production environment.

Another option in controlled atmosphere plasma-spraying is spraying in an underwater environment.[54] This method imparts a rapid quench that can be advantageous for some coating materials. In addition, underwater spraying controls noise, eliminates airborne dust, and facilitates collection of overspray.

Chamber spraying has also been utilized to synthesize novel compounds by reaction of the in-flight droplets with the atmosphere. Titanium nitride coatings have been generated by plasma-spraying in nitrogen at 2 atm (0.2 MPa) pressure.[55] Another example is production of titanium aluminide powders and free-standing shapes by plasma-spraying of aluminum in a titanium chloride atmosphere.[56]

Other Refinements in Plasma-Spraying

The United States Navy has developed its own wire fed plasma-spray system which is claimed to provide bond strength and coating densities not previously attainable with wire fed systems.[57]

Coatings applied by this technology include alloy 625 (UNS N06625), Monel† (alloy 400, UNS N04400), nickel-aluminum bronze, type 420 (UNS S42000) stainless, and carbon steel.

Inductively coupled plasma is emerging as an alternative to conventional plasma with some attractive features.[58-60] Because there is no cathode or anode, electrode contamination is eliminated. Almost any gas can be used, an advantage in maintaining stoichiometry of some oxides and possibly inducing desirable gas-feed reactions (e.g., nitride formation) during spraying. Axial injection of feedstock and long feedstock residence time in the heating zone diminish the likelihood of defects associated with unmelted particles. Larger particles can be sprayed, which reduces feed cost, increases deposition efficiency, and leads to less oxide in the coating when metals are sprayed. This last factor is especially important for oxygen-sensitive refractory metals such as tungsten and titanium aluminides.[60]

Another innovation in plasma technology is the "water stabilized" plasma-spray system.[61] In this type of plasma torch, the arc is established between a graphite rod cathode and a rotary metal anode. A peripheral swirl of water is used to stabilize the arc. The oxygen-hydrogen plasma from water decomposition in the arc is characterized by very high temperature and thermal conductivity. This allows high feed throughput on the order of 40 kg/h (88 lb/h) for aluminum oxide. Because of the oxidizing nature of the plasma, water stabilization is confined to spraying of oxides. However, other fluids can also be used as the stabilizing medium, so a wide range of materials can be sprayed. In the United States, Minnesota-based Plasmafusion is active in this area.[62]

The Plazjet-II† 200 system is a hypersonic plasma-spray system utilizing extremely high power input (200 kW). The system is claimed to offer advantageous high deposition rates and high energy efficiency. Unlike conventional HVOF, ceramics can

† Trade name.

be effectively sprayed. The technology is marketed in the United States by Whitco.

Another high-energy system is the Gator-Guard† technology developed by United Technologies for aerospace applications and now proprietary to Sermatech. Details on the process are not available, but a gas shroud is apparently used to accelerate particles in flight and control gas-particle reactions.[63] Tucker provides data on an argon-shrouded plasma-spray process.[36] Titanium sprayed without shrouding had an oxygen content greater than two percent, while the shrouded system yielded coatings with 0.7 percent oxygen, nearly the same as that of the starting powder.

Electric Arc-Spraying with Inert-Gas Atomization

As a low cost, high production rate alternative to chamber spraying, arc-spraying may be performed using a nonreactive gas such as nitrogen or argon to atomize the dual wire feedstock melted by the arc. Kaiser and Miller have documented the improvement in coating quality attainable by this approach.[33] Inert-gas atomization reduces the detrimental effects of metal oxidation that occurs largely in the vicinity of the arc zone. Figures 5.1 and 5.2 indicate the resultant improvement in coating microstructure and oxide content for arc-sprayed Ni-Cr and type 420 stainless steel. For arc-spraying of aluminum, inert-gas atomization results in a significant improvement in deposit efficiency (Figure 5.3).

In the case of arc-sprayed high-carbon steels used for wear reclamation, carbon loss as high as 42 percent can be experienced even if power input is as low as possible.[29] Using nitrogen as the atomizing gas alleviates this problem considerably and significantly upgrades coating service life.

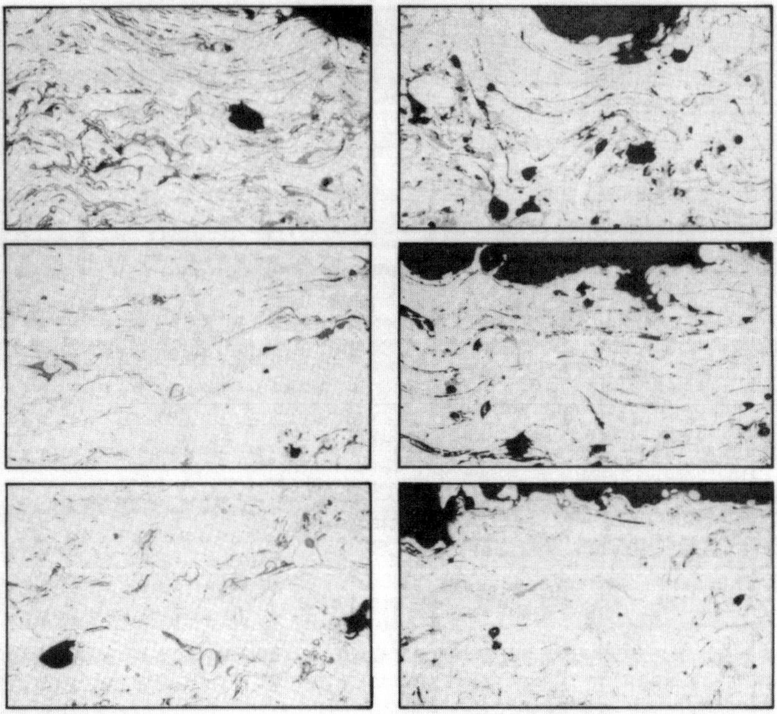

Figure 5.1 Coating micrographs indicating the benefits of arc-spraying with inert gas as the atomizing medium. Nickel-45% chromium (left column); type 420 stainless (right column). Inclusions (gray) and oxide striations between splats are markedly diminished in going from air atomization (top) to nitrogen atomization (middle) to argon atomization (bottom). Isolated porosity (black) persists in the inert gas-atomized product. 200X, unetched.[33] Used with permission of ASM International.

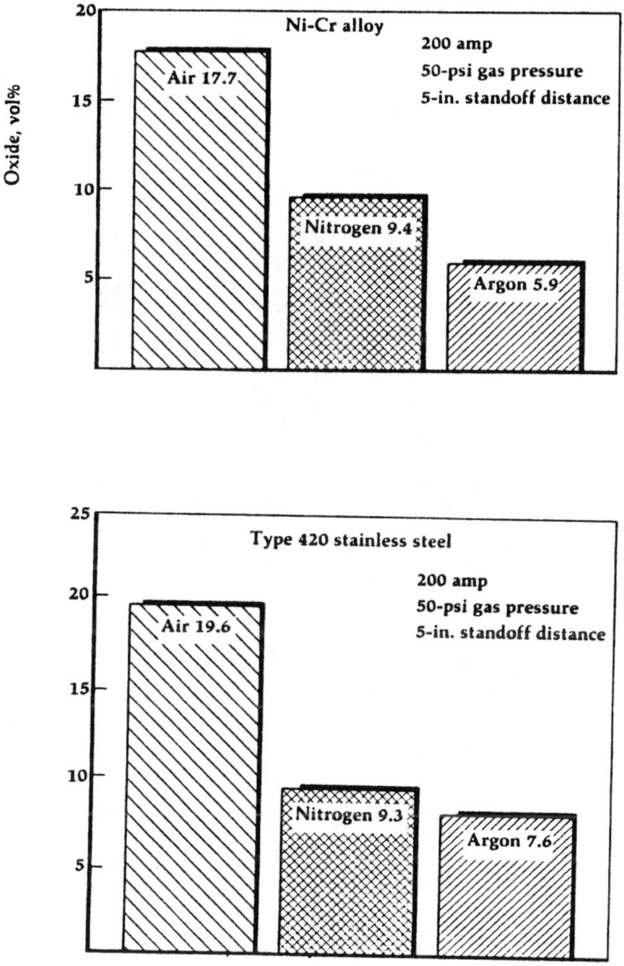

Figure 5.2 Oxide content of the coatings in Figure 5.1 quantified by image analysis. Oxidation is reduced by the inert-gas atomization, which removes oxygen from the vicinity of the arc zone.[33] Used with permission of ASM International.

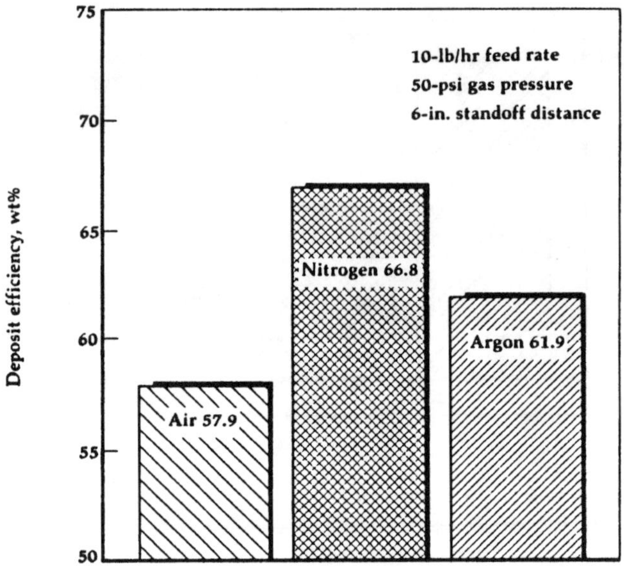

Figure 5.3 Inert-gas atomization significantly improves the deposit efficiency of arc-sprayed aluminum on low-carbon steel. This relates to (1) reduced oxidation in the arc and spray stream, (2) lower arc temperature, which reduces both vaporization and oxidation losses, and (3) thinner oxide shells on large droplets, increasing the likelihood that they will adhere to the workpiece.[33] Used with permission of ASM International.

Other Refinements of the Electric Arc Process

Marantz of Flame-Spray Industries has developed a single wire fed "arc-plasma" system in which a plasma transferred arc is formed between the cathode and feed wire anode.[64] Coating quality is reportedly comparable to that of powder-fed air plasma-spray, while the advantages of the wire arc process are retained. Extremely high deposit efficiencies and throughputs have been reported. The system is especially appropriate for oxidizable feeds like molybdenum, aluminum, and aluminum-alumina composite wire because the inert plasma gas acts as a shroud to inhibit oxidation during flight.

Marantz is also responsible for a liquid metal arc process, envisioned as an alternative to galvanizing for in-plant coating of large surface areas of steel.[65]

Another noteworthy innovation in arc-spraying is the Spraytec system, which utilizes a cored wire feed usually involving carbides for wear resistance.[66] The key advantage over HVOF or plasma is a much higher deposition rate, especially important for coating large industrial rolls and wear plates.

Dallaire and Levert have described a "reactive core wire" arc-spraying approach in which ferro-titanium, boron, and iron powders are sheathed in carbon steel or type 304 (UNS S30400) stainless.[67] The resultant composite coating contains a significant volume fraction of titanium boride, which is advantageous in terms of wear response.

HVOF Innovations

HVOF spraying is the basis for more advanced systems yielding premium metallic coatings for corrosion protection. A primary example is the Ultracoat† technology developed by Amoco and currently being licensed to commercial applicators.[68] The objective of the Amoco program was to attain wrought-quality, nonporous coatings for corrosion protection in various chemical and petrochemical applications without resorting to controlled atmosphere chamber spraying. Ultracoat is based on a commercial HVOF torch with hydrogen used as the fuel gas. The key to the innovation is a patented inert gas shield used to isolate the feed-carrying flame from the ambient atmosphere.[69]

Ultracoat has been extensively evaluated by Amoco under controlled laboratory conditions and in field service. Details will be provided in the chapters on applications. Most work to date has been with Hastelloy C and type 316L (UNS S31603) stainless steel. Figure 5.4 compares coating microstructures for Hastelloy C applied by Ultracoat and a variety of alternative systems. The marked reduction in intersplat porosity and oxide is critical to acceptable performance in chemical and petrochemical process equipment subject to corrosion problems. Very thick (60 mil [1.5 mm]) Ultracoat coatings can be attained without residual stress problems, which is a further advantage in establishing a reliable corrosion barrier. Ultracoat spraying is relatively insensitive to spray angle, an important consideration for complex geometries like boiler tube waterwall panels.

Similar HVOF developments have apparently been pursued by other companies (e.g., Metalspray Inc., Richmond, VA), but details are not available.

Another major commercial innovation based on HVOF is the Browning Aerospray† High Velocity Air-Fuel (HVAF) system.

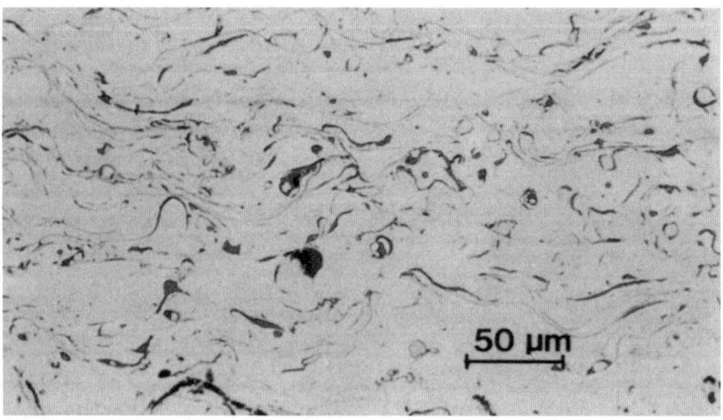

Figure 5.4a Hastelloy C coatings. Air Plasma from commercial vendor. Courtesy of Amoco.

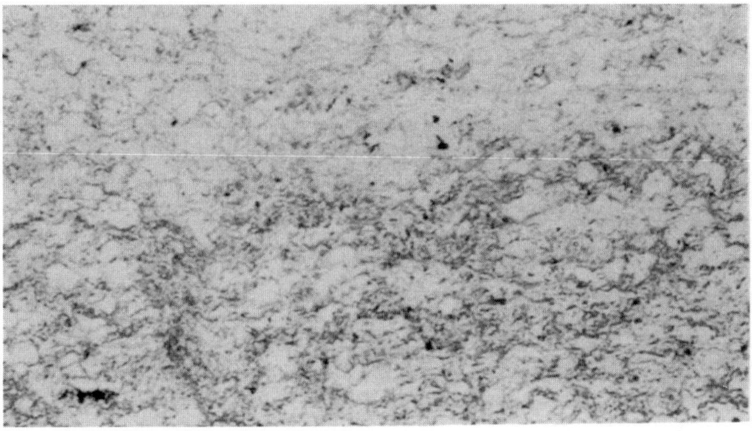

Figure 5.4b Hastelloy C coatings. Jet Kote HVOF from commercial vendor. Courtesy of Amoco.

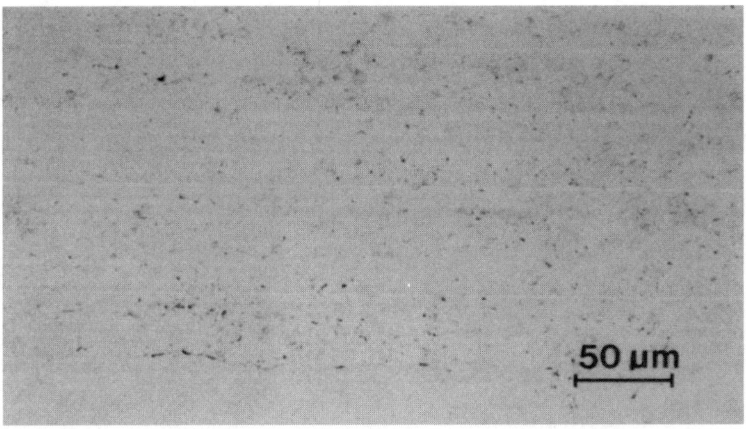

Figure 5.4c Hastelloy C coatings. Amoco Ultracoat Process, which yields reduced porosity and entrapped oxide. Courtesy of Amoco.

Air, rather than oxygen, sustains the combustion, and this same air flow is used instead of water to cool the gun. The result is a marked improvement in process economics while the quality of HVOF is still realized.

Most recently, Browning has introduced the Hypersonic Velocity Impact Fusion (HVIF) approach.[70] The temperature of the expanded combustion gases is held low enough to avoid melting of even the smallest particles in the feedstock. Conversion of kinetic energy to heat on impact is relied upon to fuse the coating, yielding low porosity and minimal trapped oxide. The technique is particularly suited to cermet coatings like tungsten carbide/cobalt.

Automation and Process Control

The Amoco Ultracoat development provides a good example of another key aspect of the maturation of thermal spray technology: the use of automation and other refinements in process control to ensure optimized coating quality.[68] One of the applications for Ultracoat at Amoco is repair of the interiors of large process vessels on site. A high-quality, impervious coating for corrosion protection requires precise control of gun-to-substrate distance and angle of incidence. Thus, hand-held spraying leaves much to be desired, particularly in the operator-unfriendly interior of a chemical process tank!

To meet this need, Amoco developed a unique microprocessor-controlled spray gun handling system, which is capable of coating the inside surface of cylindrical vessels ranging from 7 to 16 ft (2 to 5 m) in diameter. Another potential application for Ultracoat is in the protection of boiler tube walls, particularly those in the pulp and paper industry's black liquor recovery boilers. This led to the development of a prototype automated gun-positioning device for coating these surfaces. Automated control of the spray gun and substrate movement is also important for chamber spraying and large-scale production of components.

Control of position is only one aspect of process control to obtain optimal coatings. Herman has observed that the reproducible attainment of target coating microstructure is a major reason for the current proliferation of thermal spray technology.[51] Mass-flow control of process gases enables digital output with feedback potential. For powder-fed systems, volumetric delivery systems provide tight control of feed rate,[71] and fluidized bed powder feeders provide very smooth delivery of feedstock. For plasma systems, advances in control of gun-related parameters include constant energy levels using thyristor power sources and continuous automatic control of the cathode-anode spacing.[51]

Process control and automation are receiving ever-increasing attention from thermal spray researchers. Workers such as Lee and Safai[72] and Mayr[73] have described sophisticated concepts such as automated monitoring of coating thickness, particle velocity and flux density, closed-loop controlled expert systems, and in-process QC (quality control) inspection and diagnostics of plasma-spray coatings. On the lower-tech end of the process control spectrum, Tucker[63] and Hermanek[20] have pointed out that problems in coating quality have often been traced to a simple lack of calibration of process instrumentation. Greater appreciation of the relationships between process parameters, coating microstructure, and performance on the part of applicators will help to ensure more reliable high-performance coatings.

CHAPTER 6

Feedstock Materials

Numerous feedstock materials are commercially available for powder spraying. However, powders have largely been developed to cater to aerospace markets, because these have been commercially attractive and allowed significant development costs to be recovered. Some of the key alloy compositions appropriate for corrosion resistance in specific chemical process environments are not now offered by suppliers of consumables. This is understandable considering the limited current use of thermal spray in such applications. As spray system capabilities advance to the point where porosity is not the limiting factor, more specialized feeds for chemical process applications will hopefully become available from suppliers. In the interim, facilities are available for pilot production of experimental compositions, which might be an option for large CPI companies wishing to explore new coatings in-house.

In considering available powder feeds, it is important to recognize that composition is not the only important variable. Particle shape, homogeneity, and microstructure depend on the specific powder manufacturing process; and these variables can have profound influence on the spraying process and the quality of the coating. In some cases, the powder may be a true composite—for instance, a coating deposited on a core of different composition. Simple blends of powders are also sprayed and may provide specific advantages. Particle size range and distribution are also important. A tight distribution is advantageous in terms of control of particle melting but results in increased feedstock cost. The emerging HVOF processes generally require a finer particle size range that may not be commercially available for many of the feed materials specified for plasma-spraying. This situation may change as HVOF becomes more widely utilized.

Availability of wire feeds is essentially restricted to ductile metal matrix materials. Besides pure metals and metallic alloys, these may be composites containing dispersions of brittle nonmetals.

Another emerging approach is the cored wire with a brittle material such as tungsten carbide sheathed in a ductile metal. The unique Rokide† ceramic spray system developed by Norton has an associated family of oxide ceramic rod feed materials.

It is also important to recognize that compositions performing well in wrought or cast form may not give service comparable to thermal spray coatings. Even if porosity is minimized, other microstructural effects may come into play, especially in corrosive environments. On the other hand, the rapid solidification associated with thermal spray may, in some cases, yield a coating with properties superior to its cast or wrought counterpart.

Powder Production Methods

Figure 6.1 gives an indication of the range of particle morphologies associated with different powder manufacturing techniques. Some of the innovative processes can add significantly to feedstock cost. For the common metallic materials, atomization from the liquid is the standard production method. Gas atomization generally yields a spherical or elongated droplet morphology, while water-atomized powder tends to be lumpy and irregular in shape. The optimal spraying parameters for gas- versus water-atomized powder can be quite different.

Figure 6.1 Representative powder morphologies for thermal spraying: (a) inert-gas atomized, spherical NiCoCrAlY powder (-325 mesh) for low-pressure plasma-sprayed coating for protection against hot-gas corrosion or bond coat for thermal barrier coatings (TBC); (b) water-atomized, irregular Ni-5%Al powder for air plasma-sprayed bond coats; (c) fused and crushed ZrO_2-7% Y_2O_3 optimally stabilized for TBC; (d) agglomerated, sintered ZrO_2-7% Y_2O_3 with spherical, porous particles fully alloyed and highly stabilized; (e) dense, sintered WC-12% Co powder for air plasma-sprayed wear coatings; (f) agglomerated, sintered WC-12% Co powder with spherical shape and porous structure for air plasma-sprayed wear coatings.[21] Used with permission of ASM International.

Kubel indicates that if water-atomized material is used with parameters adjusted for gas-atomized powder, problems with nozzle clogging, coating discoloration, cracking, and spalling may ensue. If a spraying campaign is conducted with parameters adjusted for water-atomized powder, then poor deposition efficiency, high porosity, and inferior bond strength may result if gas-atomized powder is substituted. For gas atomization, reaction with the atmosphere in the atomizing chamber may be a consideration.

For instance, titanium and tantalum may be degraded by pick-up of oxygen and nitrogen, demanding use of an inert atmosphere.[30] The solidification rate of some alloy powders can have a great influence on the chemical homogeneity of the powder, and this can be affected by atomization atmosphere, superheat, and other aspects of the atomization process. Microsegregation of the powder will influence remelting characteristics, which may be reflected in the quality of the coating. On-line control of powder particle size is a focus of current research and development relating to gas atomization, and the fruit of this will be more economical powders of appropriate size range.[74] Another example of innovation in atomization is the Ames Laboratory high-pressure gas atomizer, which yields a very fine particle size and rapidly solidified microstructure.[75]

Ceramics, carbides, and brittle alloys may be processed to powder by casting, reactively sintering, or producing the desired composition by a chemical process and then crushing the bulk material. Crushing tends to yield powder with variable shape. Segregation in the starting material will lead to a significant composition range for the powder.

Some chemical processes yield very fine powders, which may be agglomerated with an organic binder by spray drying to yield a larger, free-flowing particle. The binder burns off during spraying, and the fine particles may coalesce and partially alloy in the flame.[76] Such powders have high surface area, low density, and melting behavior of the fine particles. Agglomeration can be used to blend material such as tungsten carbide and cobalt. The agglomerated powder can be densified by a subsequent sintering step.

Plasma has been utilized as a heat source in a number of innovative powder production methods involving sintering, reaction with ambient gases, or generation of unique powder morphologies. For instance, Metco hollow spherical (HOSP) ceramic powders based on aluminum-, zirconium-, and chromium-oxides produced by plasma fusion offer steady feed rates and consistent melting behavior, which correlates with improved deposit efficiency compared to cast and crushed materials.[76] GTE-Sylvania offers an array of "plasma densified" powders processed by agglomeration of fine powder mixes, plasma melting, and rapid gas quenching.

True composite powders have been produced by various methods, including electrochemical or electroless deposition, organic bonding, sintering, and gas phase reaction. In many instances, particles will not homogenize during the spraying process. A particle with a surface of radically different chemistry may behave quite differently from a sprayed prealloy feed with respect to adhesion, cohesion, and in-flight oxidation. This has been researched extensively with respect to composite vs prealloyed nickel-aluminum bond coats.[5,9,23] Another composite approach is the addition of fibers or whiskers to the powder feed to improve coating mechanical properties.[77,78] This development is still at the research stage.

A common theme in the recent technical literature on thermal spray has been the comparison of coating quality using different feed forms of the same composition. Materials studied have included tungsten carbide cermets,[79-81] partially stabilized zirconia,[82-83] titanium,[30] and the nickel-aluminum bond coats mentioned above.

Iron-Based Alloys

A variety of low- and high-carbon steels, cast irons, and stainless steels are available in powder and/or wire form for part restoration, enhanced wear performance, and corrosion protection. Stainless compositions available are equivalent to wrought grades 202, 304, 316, 420, and 431 (UNS S20200, S30400, S31600, S42000, and S43100, respectively); although it needs to be stressed that, independent of porosity issues, thermal-sprayed coating microstructure and performance may not be the same as for the corresponding wrought grades. There are also some unique stainless formulations, such as Stellite Tristelle† powders modified with cobalt and silicon (Si), and Miller Thermal Tufton† wire formulations with up to 1% C and 2% Cr for abrasion resistance. Tristelle compositions form chromium-rich carbides and offer enhanced corrosion and wear resistance. They are amenable to plasma-spray, HVOF, and plasma transferred arc processing.

Boiler protection, particularly for black liquor recovery boilers used in pulp and paper manufacturing, has prompted development of some specific iron-based feeds. These include Kanthal† M wire (Fe-22% Cr-6% Al) and Metco composite powders for plasma-spray, including Fe-Cr-Al-Mo and Fe-Cr or Ni-Cr composited with aluminum and yttria.

With reference to austenitic stainless steel compositions, it is noteworthy that powders and sprayed coatings may display some ferritic character.[68, 84] For fully dense coatings, such fine nuances of structure could exert a controlling influence on corrosion performance.

Metco and Sulzer Plasma Technik appear to be industry leaders in the development of composite powder materials for enhanced bond strength and performance. For example, Metco 449P powder has a cast-iron core clad with aluminum, molybdenum, boron, and iron oxide, and is designed to give a hard, low-shrink coating that is machinable to an exceedingly fine finish and appropriate for buildup of ferrous substrates. Other Metco composites include 444 (self-bonding stainless), 463 (high-chromium stainless), and 350C (high-carbon iron plus molybdenum).

Metco Diamalloy† 1008 is a blend of two iron-based powders with particle size tailored for HVOF spraying. The resultant coating is high in chromium and molybdenum and is intended for corrosive wear applications below 1,200°F (650°C). It is suggested as an economical alternative to chrome plating.

Another family of innovative iron-based powders is offered by Wear Management Services.[85] Tribolite† is a cobalt-free formulation offering outstanding resistance to adhesive wear and galling, good thermal stability, and attainment of an exceedingly fine surface finish by machining. Wear performance is apparently related to the presence of fine boron-rich phases in the microstructure. Modification of the powder formulation yields feedstocks tailored for abrasion resistance and restoration of steel part dimensions. By a proprietary blending with bond-enhancing powder, these materials can be modified to one-step feeds sprayable by simple combustion torches without the need for a bond coat. This feature is especially attractive for low-cost coverage of large surface areas. Compositions are also available for spray-and-fuse and plasma transferred arc application.

Also noteworthy among the iron-based, wear-resistant coatings are the "metamorphic alloys" promoted by Amorphous Metals Technologies.[86] It is claimed that abrasion in service transforms the coating surface to an amorphous state, rendering it

† Trade name.

more resistant to abrasive wear. One member of this product line, Armacor C,[†] is suggested for various CPI and pulp and paper applications because of superior resistance to bleach, chlorides, and medium-strength sulfuric acid.

The recent technical literature includes evaluations of "cored wire" feeds—low-carbon steel sheaths with powder cores of rare earths, hard oxides, high-carbon ferrochrome, or chromium carbide.[87, 88]

Iron aluminide intermetallic powder, produced by Ametek Specialty Metals, is intended largely for fabrication into components via powder metallurgy techniques. However, outstanding resistance to high-temperature oxidation, sulfidation, and molten salt attack makes it an attractive candidate for coatings in several CPI applications.

Zinc- and Aluminum-Based Feeds

Pure zinc and aluminum are the dominant thermal spray coating options for corrosion protection of structural steel. Aluminum is preferred for severe marine applications, some acidic industrial atmospheres, and cases of elevated temperature exposure. Zinc is the workhorse coating for other applications, because it is more forgiving than aluminum with respect to surface preparation. Because deposition rate is important for large surface area corrosion protection, wire-fed combustion and (especially) arc-spray processes are dominant. However, Eutectic and Metco offer powder combustion systems capable of fairly high deposition rates.

Zinc-15% aluminum wire is another option and offers extended life compared to zinc in severe environments while avoiding aluminum's surface preparation sensitivity and lack of sacrificial action in nonchloride exposures. This European innovation has not been extensively practiced in the United States, possibly because of a negative assessment by the U.S. Naval Research Laboratory.[89] Zinc-20% tin, a zinc-based solder composition, is also available in wire form.

A more recent innovation is the Duralcan† 90/10 (aluminum-10% aluminum oxide) composite wire developed by Alcan. This feed offers the spraying characteristics and corrosion response of aluminum, with the oxide present to enhance wear, abrasion, and antiskid properties (Figure 6.2).

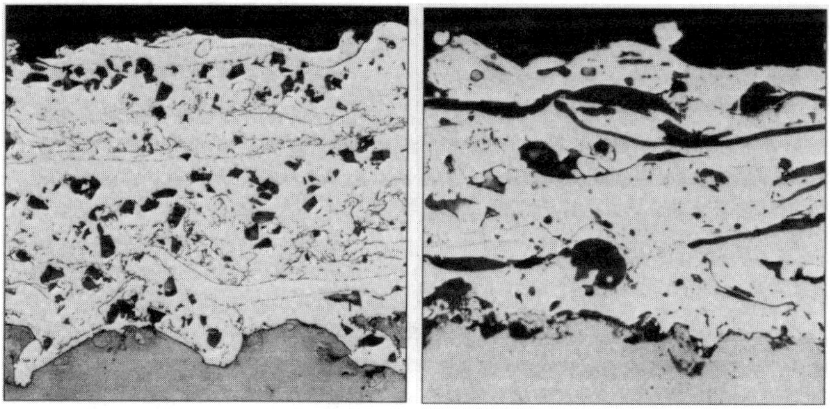

Figure 6.2 Microstructures of Duralcan Al-10%Al_2O_3 coatings applied by wire flame-spraying (left) and arc-spraying (right); 200X. Partial melting of the oxide component is visible in the arc-sprayed coating. This diminishes the enhancement of abrasion resistance due to the oxide phase, although other aspects of wear performance are improved. From Duralcan product literature; used with permission.

Innovative zinc-based powders have recently been explored for a variety of corrosion-related applications. Zinc-30% nickel produced by casting and crushing has been assessed as a corrosion-resistant bond coat for plasma-sprayed oxides,[90] and powder with nickel content on the order of 15 percent has been assessed as a feedstock for combustion spraying of corrosion-resistant coatings on steel.[91] Intermet Technology has evaluated a zinc-silicate composite powder combustion spray for corrosion protection of steel in severe environments, including reinforcement in chloride-laden concrete.[92]

Besides pure aluminum wire, Miller Thermal provides an aluminum-based wire with 4.5% to 6.0% Si. This material sprays faster and machines easier than pure aluminum and produces a finer surface finish, but its corrosion performance is not as good. Aluminum-12% silicon, the lowest melting eutectic composition of this alloy system, is available in both powder and wire form. Aluminum and aluminum-silicon are the basis for abradable composite powders containing graphite or polyester.

Copper-Based and Miscellaneous Nonferrous Metal Feeds

Readily available copper-based wire feeds include pure copper, brass (copper-zinc) of various compositions, copper-nickel, and several types of bronze. Spraying of aluminum-bronze wire of composition Cu-9% Al-1% Fe provides an excellent machined surface and is extensively utilized for soft bearing surfaces. Aluminum-bronze is also available in powder form, as are pure copper and copper-nickel. Alloying copper-nickel powder with indium provides a very dense coating resistant to fretting and galling. Aluminum-bronze is the basis for composite and blended powders containing polyester and boron nitride for abradable coatings. Hobart-TAFA has reported on a novel cored silicon-bronze wire containing aluminum and tin that is used in an automotive application as a replacement for lead-based solder. Copper-sheathed silicon carbide is also available. The U.S. Department of Energy Ames Laboratory has developed a copper-chromium spray feed for protection of copper in highly oxidizing environments such as rocket thrust chambers and nozzles.[75]

Tin-based wire feeds include pure tin and Sn-7% Sb-3% Cu "babbitt" for soft bearing applications. Lead and lead-tin solder have been sprayed in the past, but environmental concerns have probably diminished their use. Other low-melting alloys are useful for moldmaking applications.

Nickel-Based Alloys

Nickel-aluminum materials, in both composite and prealloyed variants, are available in powder and wire form and are the most common bond coats for ceramic overlayers. Standard compositions are 5% and 20% aluminum. Nickel is also one of the possible components for some of the "M-CrAlY" compositions useful for hot corrosion and as oxidation-resistant bond coats for thermal barrier coatings. Bond coats are covered in depth in a later chapter. Nickel powder composited with aluminum and molybdenum is a useful "self-bonding" coating for reclamation work.

Other wire feeds include pure nickel, alloy 400 (UNS N04400) with approximately 30% Cu, nickel-chromium alloys 80/20 and 60/16 (with balance Fe), and alloy 625 (UNS N06625), with 20% to 23% Cr and significant molybdenum and niobium content. TAFA has recently promoted use of TAFALOY-45CT† (with 44% Cr and 1% Ti) for boiler applications.

Along with pure nickel, a broad array of nickel-based alloy and composite powders are available. These include binary alloys with 20% and 50% Cr; alloys 600 (N06600), 625, and 718 (N07718); and various aerospace superalloys. The Alloys International division of Miller Thermal offers a composite of 80/20 Ni-Cr coated with aluminum that gives enhanced bonding and a self-bonding powder with 8% Cr, 7% Al, 5% Fe, 5% Mo and provides an attractive combination of machinability, wear and corrosion resistance, and high-temperature oxidation resistance. Sulzer Plasma Technik offers Ni-19% Cr-6% Al in prealloy and composite form. Anval is licensed by Haynes International to make powders of its alloy C-22† (N06022) alloy for use with thermal spray alloys. Stellite offers eleven different powders akin to the Haynes International Hastelloy† alloys under its Deloro-Nistelle label, including alloys C, C-22, C-276, and D. Metco offers several composite powders based on Ni-Cr and Ni-Fe, with added aluminum and molybdenum.

Another recent Metco development is the 700/700F family of nickel-based powders for "amorphous-like" coatings. Stellite offers a nickel-based Tribaloy† T-700 powder with 32.5% molybdenum that provides outstanding wear resistance because of the presence of molybdenum-rich intermetallic compounds. See the cobalt subsection for more on Tribaloy alloys.

"Self-fluxing" nickel-based powders are used extensively in "spray-and-fuse" applications. Alloys typically contain 10% to 20% Cr, and 2% to 4% each of Si and B. Key suppliers include Stellite, Eutectic, Wall Colmonoy, UPT, and Metco. Colmonoy literature emphasizes the importance of chromium boride crystals in imparting wear resistance to coatings like Colmonoy No. 6. Colmonoy has also developed new higher-chromium formulations (Nos. 8 and 84) intended to substitute for cobalt-based spray-and-fuse compositions in high-temperature service environments. Another noteworthy composition is 69SC with 5% Mo for enhanced corrosion response.

The American Welding Society (AWS)[(2)] handbook on thermal spraying indicates the availability of cord (i.e., filled plastic tube) feeds of nickel-based self-fluxing alloys, but cord spraying is apparently practiced almost exclusively in Europe.[38] The same type of alloy is available for arc-spraying in the form of powder in a nickel sheath.[93]

(2) American Welding Society, Miami, FL.

Nickel-titanium has reportedly been used as the bond coat in conjunction with a chromium top coat for black liquor recovery boilers.[94] Other unusual nickel-based feeds include abradable composites containing graphite and boron nitride. Ametek has commercialized nickel aluminide intermetallic powder with boron added for improved ductility. Although the main focus is on powder metallurgy applications, the high-temperature oxidation resistance of this material suggests its use as a coating feedstock.

Cobalt-Based Alloys

Cobalt-based formulations offer an attractive combination of high-temperature corrosion and oxidation resistance, along with superior wear performance. The Stellite[†] family of cobalt-chromium alloys is well known, and Stellite Coatings offers several powder formulations appropriate for plasma-spray, HVOF, spray-and-fuse, and plasma transferred arc. Most of these contain appreciable amounts of tungsten. An exception is Stellite 21, which is tungsten-free and high in molybdenum.

Another cobalt-based alloy with high molybdenum is the Haynes International Ultimet[†] alloy produced in powder form by Anval. Its composition was developed to optimize both corrosion and wear resistance. Metco offers a self-fluxing (i.e., spray-and-fuse) powder with 40% Co, 27% Ni, 18% Cr, and 6% Mo, along with silicon and boron. A cobalt-based self-fluxing cord feed is also reportedly available.[38]

Another premium wear-resistant coating is based on the Tribaloy family of Co-Mo-Cr-Si alloys developed by the DuPont Company. Cobalt-molybdenum-silicon compounds in a cobalt alloy matrix impart outstanding resistance to adhesive wear, a low coefficient of friction against steels and superalloys, and good oxidation resistance up to 1,000°C (1,830°F). Cobalt is also a major constituent of some of the "M-CrAlY" formulations for high-temperature thermal barrier bondcoats.

Carbide and Boride Feeds

Hard carbide phases are the basis for many of the premium wear-resistant coating formulations. Typically, the carbides are held in a tough metallic matrix, so the coatings are termed "cermets" or cemented carbides. Powders can be produced by various methods: cast or dense sintered and crushed; agglomerated; agglomerated and sintered; clad or otherwise coated. A recent study compared the various types of tungsten carbide-cobalt powders for plasma-spraying with respect to carbon loss, porosity, and wear performance.[81] A key consideration is retention of the desirable WC phase in the final coating. Sometimes, the carbide-containing powder is blended with other powders to provide such features as self-fluxing or enhanced bonding.

The most prevalent type of cermet is based on tungsten carbide. For instance, Miller Thermal offers 26 different tungsten carbide powders and blends. The most common cementing matrix metal is cobalt in the range 7% to 20%. Higher cobalt improves impact resistance at the expense of abrasion resistance. The tungsten carbide/cobalt powder may be blended with a nickel- or cobalt-based self-fluxing alloy for spray-and-fuse applications. Additionally, nickel-aluminum may be added for enhanced bonding. Other tungsten carbide variants include a blend with chromium carbide and nickel (Miller Thermal AI-1085) specifically suggested in the product literature for demanding high-temperature plus corrosion environments in the chemical industries. GTE-Sylvania offers a number of novel powders based on tungsten carbide. WC-10% Co-4% Cr ("GTE SX-178") provides better oxidation resistance than WC-Co formulations and is cited for applications such as oil field gate valves, paper rolls, and certain CPI environments. WC-TiC-17% Ni ("GTE SX-304") is described as having particularly good resistance to alkaline solutions, combined with resistance to erosive wear.

Another group of carbide cermet powders available from several suppliers utilizes chromium carbide (Cr_2C_3) in a nickel-chromium matrix. This provides good resistance to high-temperature abrasive, fretting, and percussive wear.[41] Sulzer Plasma Technik offers a novel three-part blend (AMDRY† 343) of chromium carbide, nickel-aluminum, and cobalt alloy designed to produce dense, self-fused coatings for high-temperature wear. "GTE-Sylvania SX-199" contains both tungsten carbide and chromium carbide in a nickel-chromium matrix and is cited for use in seals and bearings for process fluid pumps. Metco offers a composite powder of chromium carbide and nickel (83VF-NS). Included in Colmonoy's product line is an alloy that evolves a novel microstructure of bimetallic carbides in a nickel-tungsten-chromium matrix (Colmonoy 88). Different powder grades are appropriate for spray-and-fuse, HVOF, and plasma transferred arc application. Chromium carbide is also available without other additions and apparently is used in some aerospace applications.

Titanium carbide is the basis for the ResisTic† family of composite powders promoted by Alloy Technology International. Options for the metallic matrix include nickel-chromium, stainless steel, and chromium-molybdenum tool steel. The powders are manufactured by GTE-Sylvania's agglomeration/plasma melt/rapid solidification process. Amorphous Metals Technologies Duocor† wire feedstock for arc-spray contains tungsten and titanium carbides in a "hard alloy" matrix.

Montreal Carbide Co. offers novel boride-reinforced stainless steel powders promoted as low-cost alternatives to WC-cobalt for abrasion/erosion resistance. Another unique feedstock is Stellite 967, which apparently consists of silicon carbide in a cobalt matrix. Silicon carbide has also been explored as a reinforcement for aluminum wire as part of the Alcan Duralcan[†] product development.[95] Compared to aluminum oxide, the silicon carbide apparently degrades corrosion performance in marine environments. Pure boron carbide reportedly can be applied by vacuum or underwater plasma-spray and may be blended with nickel-chromium or other nickel-based alloys.[96]

Refractory Metals

The extraordinary corrosion resistance of the refractory metals (titanium, chromium, tantalum, molybdenum, niobium, tungsten, and zirconium) implies a great deal of usefulness in the chemical process industries. However, with the exception of molybdenum, use to date has not been widespread, primarily because of the problem of reaction with the ambient atmosphere during spraying. With the advent of controlled atmosphere spraying, some of the potential for utilization of these materials will undoubtedly be realized.

Molybdenum-based materials are used extensively in spraying bearings and automotive piston rings because of good resistance to adhesive wear. Pure molybdenum wire is available in both black and chemically cleaned forms. Flame-sprayed wire reportedly gives wear performance superior to that of plasma-sprayed powder.[34] Molybdenum-based powders are not generally made by atomization because of gas interaction problems. GTE-Sylvania is a major supplier of powder, with options including pure molybdenum, self-fluxing nickel alloy blends, and composites of Mo/Mo_2C and Mo/MoO_2. Metco offers a high-carbon iron-molybdenum composite powder. Mo-25% Ni is an advantageous plasma-spray feed for the resurfacing of Yankee drier rolls. Pure molybdenum is available in the proper size range for HVOF spraying. Sulzer Plasma Technik has a new prealloy powder (SP-

174) with approximately 40% Mo for wear-related applications.

Titanium, chromium, tantalum, and tungsten are available in powder form. Watson Grinding and Manufacturing has commercially sprayed a Nb-30% Ti-20% W alloy. Research has indicated the feasibility of chamber arc-spraying titanium and tantalum,[30] and titanium has been HVOF sprayed on nickel-based alloys for prevention of crevice corrosion.[97] Zirconium has been successfully sprayed by wire-arc with inert gas atomization, and vacuum arc spraying has been explored.[88] This may eventually lead to some significant commercial use of wire feeds.

Ceramics

Ceramic coatings offer many attractive features, including wear resistance, chemical stability, low thermal conductivity for thermal barrier applications, and electrical insulation. The main drawback is a lack of ductility, and various strategies are pursued to inhibit coating fracture.

Workhorse oxide powder feeds include chromium oxide (high abrasion resistance and environmental stability), aluminum oxide with various impurities, and alumina-titania, which can be ground and lapped to a very smooth surface finish.[41] Increased titania content improves toughness but reduces wear resistance. Straight titania exhibits outstanding toughness.[99] According to d'Angelo and El Joundi, silica additions to chromia yield the most durable ceramic wear coating, with silica functioning "as a damping medium that absorbs energy from impact loading, thus preventing grain loosening, which further enhances wear resistance."[99] Alumina-magnesia spinel has outstanding insulating properties.[24] It is noteworthy that Metco literature for its hollow spherical (HOSP) plasma-fused chromium oxide includes data on combustion spraying. Apparently, the uniform melting of the hollow particles improves spraying by this method.

Zirconium oxide-based materials are used extensively for thermal barrier applications. "Stabilized zirconia" refers to the fact that additions of calcium, magnesium, or yttrium oxide promote formation of the cubic form of zirconium oxide. Attainment of a "partially stabilized" phase structure is a mechanism for toughening structural ceramics. Tetragonal zirconia in the cubic matrix transforms to a monoclinic structure under the action of stress at the tip of a crack. The increase in volume associated with this phase change acts to inhibit further crack propagation. However, this mechanism may not be operational for zirconia coatings because the pre-existing microcrack network may be largely responsible for the enhanced mechanical response.[100]

Partially stabilized zirconium oxide powders with 7%, 8%, and 12% yttria are available. Curiously, recent research on cavitation erosion has indicated that the optimal toughening is for 2% yttria.[12] Recent technical literature provides performance comparisons of the various forms of powder, including Metco's HOSP powders[23, 82, 101]. These are claimed to offer uniform feed rates, consistent melting behavior, and superior deposit efficiency. Fully stabilized zirconia variants include 20% calcia and 20% yttria powders. The technical literature also includes reference to ceria- and ytterbia-stabilized zirconia,[100, 102] and Metco offers a 205NS powder with both ceria and yttria additions. The latter is another of Metco's HOSP materials and is claimed to surpass the cyclic thermal fatigue resistance of zirconia-8% yttria while offering lower thermal conductivity to enhance thermal barrier performance. Pure zirconium oxide, zirconia-titania-yttria, calcium zirconate, and magnesium zirconate are also available from various suppliers.

Herman has reported on the ceramic alloy "Cordierite" (2MgO-$2Al_2O_3$-$5SiO_2$), which is amorphous as-sprayed and exhibits outstanding chemical durability.[103] Surface Alloying & Wearfacing Services offers a fracture-resistant Crystallite[†] glass coating of unspecified composition with potential for use in lining reactor vessels. F.J. Brodmann offers a range of porcelain powders for thermal spraying, including blends with stainless, nickel-aluminum, silicon carbide, graphite, and boron nitride. Metco and Sulzer Plasma Technik provide blends of magnesium zirconate with nickel-aluminum, nickel-chromium, and nickel-chromium-aluminum. The Alloys International division of Miller Thermal offers aluminum oxide, chromium oxide, and Cr_2O_3-SiO_2-TiO_2 for HVOF spraying.

A broad range of ceramic rod feeds is available for use with the Norton Rokide† combustion system. These include alumina, chromia, pure and 8% yttria-stabilized zirconia, zirconium silicate, magnesium zirconate, magnesium aluminate, and alumina with 3%, 13%, and 20% titania.

Plastics

The thermal spraying of plastic materials is starting to see increased attention and should be of great interest to the CPI. Spraying is a means of applying coatings free of volatile components in a thick single coat with no overspray damage and short cure time. The process is relatively insensitive to ambient temperature, and preheat from the combustion torch removes moisture from the substrate surface. Advances in feed materials and better control of the coating process should alleviate past problems with adherence and coating degradation.[104] In particular, heat input is critical, and low-temperature guns such as UTP's UTPlast† F-311 have been specifically developed for the spraying of thermoplastics.

Brodmann is the supplier of a broad range of polymer powders, including acrylic resin EAA, nylon 11, fluoropolymer ECTFE, and three varieties of Teflon† fluoropolymer (ETFE,

FEP, and PFA). Teflon materials are also available as blends with bronze, aluminum, aluminum-silicon, boron nitride, titanium boride, and boron carbide. Teflon fluoropolymer FEP is also provided by Brodmann in wire and rod form.

The EAA resin (ethylene-acrylic acid copolymer) has been identified by workers at Dow Chemical as offering significant advantages with respect to adhesion and abrasion resistance.[105] DuPont offers a similar blend based on methacrylic acid (MAA).[106] UTP is a leading supplier of the advanced plastics, including EAA (UTPlast Extra[†]), nylon 11 (Super Plast[†]) and the low-melting ethylene-vinyl alcohol (UTPlast[†]).

McQueen of Cortech Systems identifies a number of generic polymers suitable for thermal spraying—polyethylene, nylon, ethylene vinyl acetate, polyester, and epoxy—but stresses that the powders must be considerably modified to tailor them for the combustion spray process.[104] The modified thermosetting polymers such as fusion-bonded epoxy have not yet shown the same flexibility and resistance to cathodic disbondment exhibited by coatings applied by other methods.

Metco offers polyester powder and blends with aluminum-silicon and aluminum-bronze. The blends are used for generating abradable coatings. Another novel feed is Metco 625, a stainless steel powder with additions of 2.5 percent each of epoxy and nylon 11. This is useful as a bond coat for fiber-reinforced polymers and other plastic substrates. Miller Thermal is a supplier of epoxy and nylon powders, and Sulzer Plasma Technik offers an abradable composite (AMDRY 2000) of aluminum-silicon and polyimide.

Appendix B highlights the powder feedstocks available from numerous commercial suppliers, and Appendix C is a compilation of sources of wire and rod. The list of commercially available thermal spray feedstocks is continually evolving. For a complete, up-to-date compilation, product information may be obtained from the suppliers listed in the ASM International "Thermal Spray Buyer's Guide," which is included in Appendix A. Another useful information resource is the interactive database of thermal spray consumables, processes, and applications available from MS-Card, Inc. through the CRC Press. This integrates information from 70 suppliers with 1,300 applications in 25 industries.[107]

CHAPTER 7

Bond Coats and Self-Bonding Materials

Introduction

Adhesion of most thermal spray coatings requires establishment of a suitably roughened substrate surface. The mechanism is not simply mechanical interlocking. Interaction of the impinging, spreading, and rapidly cooling spray droplets with the "anchor-tooth" profile of the surface generally determines the quality of the bond. However, some spray feeds are capable of bonding effectively to a smooth (even polished) substrate. These "self-bonding" materials may be applied alone or as thin (3 to 7 mil [0.08 to 0.2 mm]) base-coats for other exterior layers. When sprayed on a grit-blasted surface, their use makes the coating process more forgiving with respect to surface preparation, process control, and use of the simpler combustion spray systems generally associated with lower bond strength. A self-bonding material may also be used when it is not possible to properly roughen the substrate surface. An intermediate bond layer can also be useful in accommodating differential thermal stresses and impeding the propagation of cracks from a brittle surface coating to the critical coating/substrate interface.

In corrosion or high-temperature oxidation/sulfidation applications, environmental performance of the bond coat needs to be considered because the outer coating cannot be relied upon for complete protection. In some cases, the bond coat may be selected to provide the primary corrosion protection, while the porous or crack-prone exterior coating is selected for wear resistance.

Feed materials used as bond coats include various nickel-aluminum formulations, aluminum bronze, nickel-chromium, the "M-CrAlYs," molybdenum, and composites that generally have aluminum as one of the surface-bonded components.

Molybdenum and Other Refractory Metals

Historically, the first self-bonding materials were refractory metals, including molybdenum, niobium, and tantalum.[108] Effective adhesion has been associated with high particle heat and the metals' affinity for oxygen. The latter characteristic might come into play by promoting highly exothermic particle surface reactions and/or by reducing the oxide film encountered on most metal substrates. Molybdenum is still extensively used as a feed for wire combustion and plasma-spraying.

Tantalum is very sensitive to oxygen content and is not wear resistant. These factors, along with cost and high melting point, have inhibited its utilization. Tantalum is, however, exceptionally corrosion resistant in some chemical process environments, for example, strongly oxidizing or reducing conditions with a high concentration of chlorides. The recent advances in controlled atmosphere spraying suggest that tantalum could be of renewed interest, perhaps coupled with a wear-resistant oxide topcoat.

Titanium may also be useful as a bond enhancing component in sprayed alloys. Pure titanium picks significant oxygen during air plasma-spraying, but this can be controlled by inert gas shrouding.[36] A reference on black liquor recovery boilers alludes to nickel-titanium as a bond coat.[94]

Nickel-Aluminum

The original nickel-aluminum feeds were composite in nature, and the mechanism for enhanced bonding was presumed to be the exothermic reaction to form nickel-aluminide. Powder formulations still commonly employed include 20% Al (nickel clad on an aluminum core) and 5% Al (a nickel core with aluminum attached to the surface with an organic binder). A 20% Al composite wire is also available (an aluminum sheath packed with Ni-Al powder).

The recent focus has been on prealloyed 5% Al materials, including gas- and water-atomized powders, and wire for arc or combustion spraying. The bond strengths of these prealloys have been found to exceed those of the composites in some studies.[5, 108]

This calls into question the importance of the exothermic aluminide formation reaction for enhanced bonding. Recent literature emphasizes the exothermic oxidation of aluminum in flight[9] and the importance of aluminum in scavenging oxygen to improve intersplat and splat/substrate contact.[76] Studies have compared the phase structure and performance of various composite, sintered, and prealloyed nickel-aluminum powders in plasma-spraying.[9, 23, 75]

Coatings produced from clad feeds may contain free aluminum and substantial nickel oxide. High-temperature oxidation resistance is inferior for the clad coating, because of less effective protection by formation of aluminum oxide barrier films. The subtle differences in coating microstructure certainly could be important in some corrosive environments as well.

The presence of minor concentrations of other elements can influence performance of nickel-aluminum. Boron[23] and phosphorus[109] are beneficial in terms of adherence, while silicon appears to lower bond strength.[23]

Aluminum Bronze

Aluminum bronze, a copper-based alloy typically containing about 9% Al with low levels of iron and/or manganese, is another useful self-bonding material. Its high thermal conductivity makes it a good candidate for surfacing of the drying rolls for paper and textiles. It is also the preferred bond coat for copper-based substrates, which are incompatible with nickel-aluminum. The bonding behavior of aluminum-bronze again can be related to the affinity of aluminum for oxygen, which promotes exothermic oxidation of particle surfaces and removes oxide from the surface of the substrate.[110]

Nickel-Chromium and M-CrAlYs

The high-temperature oxidation resistance of nickel-20% chromium enables it to be used as a bond coat for oxide coatings in high-temperature service. The M-CrAlYs (M = Ni, Co, and/or Fe) are used extensively as bond coats for thermal barrier oxide coatings in gas turbine applications which may involve vanadate corrosion. Suppliers such as Sulzer Plasma Technik provide many different M-CrAlY formulations. Additions such as silicon, hafnium, and tantalum enhance performance in some service environments. Advanced coating systems of this kind may employ several layers, with a gradual transition from metal to oxide. These are termed "functional gradient" or "graded" coatings.

Composites and Blends

Besides binary nickel-aluminum, prevalent composite self-bonding powders include Ni-Cr-Al (for high-temperature oxidation resistance), Ni-Al-Mo (Metco 447, for tough, wear-resistant build-up coatings), and self-bonding powders with cores of low-carbon, high-carbon, and stainless steels (CEAC and Metco).

Simple blends of self-bonding powders with other feeds may also be effective in promoting adherence. For instance, Tribolite† "One-Step" offered by Wear Management Services takes this approach to provide an iron-based, wear-resistant coating that can be applied with simple combustion torches without the need for a separate bond coat.

Metco offers a novel bond coat for plastic substrates that consists of stainless steel powder with modest additions of epoxy and nylon 11. Another new bonding approach for nonmetal substrates is to coat the substrate with epoxy loaded with hollow glass microspheres. The spheres are fractured to evolve the proper surface profile for good adherence of the thermal spray coating. This is the basis of the Copperlok† technology for depositing Cu-Ni marine antibiofouling coatings on boat hulls.

† Trade name.

CHAPTER 8

Surface Preparation

A quality thermal spray coating requires proper attention to conditioning of the substrate surface prior to spraying. A roughened surface profile (anchor tooth profile) is generally necessary to ensure adequate bonding. Contaminants and blasting residues must be avoided. Oxide films cannot usually be eliminated completely, but excessive scales and corrosion products are unacceptable. Modest preheat eliminates adsorbed moisture from the surface, but too much heating can reestablish an undesirable thermal oxide. Preheating can also help to alleviate residual stresses in the coating. Masking materials are available to block coating development in selected areas.

Initial Cleaning

A wide range of cleaning procedures are available to remove surface contaminants. Some commonly practiced options include solvent or detergent washing, steam cleaning, pickling, ultrasonic methods, wet and dry abrasive blasting, and wire brushing of local areas.

Acid- or alkaline-contaminated surfaces need to be neutralized, rinsed, and dried. Effects of oil bleed-out from porous castings or dye penetrant inspection can be prevented by a baking procedure.[38] Chlorinated solvents for sensitive materials like titanium must be avoided,[38] and hydrogen introduced by any electrochemical treatment needs to be considered if the substrate is prone to hydrogen-related mechanical degradation.

Roughening by Grit Blasting

A roughened surface profile provides mechanical interlocking, increases surface area for other bonding mechanisms, breaks up oxide skins on impacting splats, and redistributes residual stresses. Roughening is desirable even if a self-bonding deposit is utilized. Roughness is characterized by a profilometer, which yields an arithmetic average of peak and valley positions relative to the mean surface plane.[35]

Recommended grit type and target profile depend on substrate and coating. Suppliers should be contacted for specific recommendations. In general, an appropriate profile will be in the range 100 to 500 microinches (2.5 to 13 µm).[38] Thicker coatings may require rougher profiles to avoid spalling. Care should be taken to avoid or remove blasting residues.

For zinc- and aluminum-based coatings, the American Welding Society (AWS) has taken a lead role in developing recommended practices to ensure coating quality.[38,111,112] An updated set of guidelines is currently in preparation. For zinc coatings up to and including 0.009 in. (0.23 mm) and aluminum coatings up to and including 0.007 in. (0.18 mm), a 300-microinch (7.6-µm) profile using SAE G-25 to G-40 aluminum oxide, silicon carbide, hardened steel grit, G-25 to G-50 silica sand, garnet, flint, or crushed slag is recommended.

For thicker coatings, a profile of 500 microinches (13 µm) using grits in the range SAE G-18 to G-25 is recommended.[38] Steel must be blast-cleaned to a white metal finish (NACE-1 or SSPC SP-5).[(3)] Maximum time delay between final blasting and spraying is 4 hours. For spraying of plastic on concrete, the

[(3)] NACE International Standard TM-0170, "Visual Standard of New Steel Airblast Cleaned with Sand Abrasive" (Houston, TX: NACE, 1970); Steel Structures Painting Counsel Surface Preparation Specification No. 5-91, White Metal Blast Cleaning" (Pittsburgh, PA: SSPC, 1991).

concrete must be fully cured and have heavy spalling and cracking areas repaired. Trowelled surfaces should be brush blasted or acid etched.[104]

Surface Machining Prior to Spraying

In parts restoration, undercutting is performed to provide a uniform base for the coating and to remove work-hardened, contaminated, corroded, or oxidized metal and previously deposited coatings.[38] A shaft should not be undercut to its end, and shoulders should have an adequate radius.

Grooving or machining of fine threads is used as a macroroughening technique to accommodate the stresses of very thick or high-shrink coatings or where the substrate has an edge. Very coarse grit blasting with SAE-14 chilled iron may also be used.

Masking

A variety of masking tapes and paint-ons are available from suppliers listed in the "Thermal Spray Buyers' Guide" in Appendix A. High impingement velocity processes such as HVOF place greater demand on masking materials. DeWal Industries has reportedly developed a new tape material for this application.[113] Evaluation criteria include ease of application, resistance to abrasive blasting, masking effectiveness (including edge retention), ease of removal, and lack of coating contamination. Higher-velocity, higher-temperature coating processes obviously place a greater demand on the masking medium.

For hard coatings, it is sometimes desirable to mask off regions of high strain. Otherwise, cracking of the coating could promote nucleation of fatigue cracks in the substrate.[63]

CHAPTER 9

Post-Treatments

Sealing Treatments

For applications such as corrosion protection, through-porosity of the coating can be a critical issue. A post-treatment that seals pores is highly desirable. Most sealers are simply low-viscosity formulations of organic resins applied by brush or spray. Air-cured options include phenolics, vinyls, polyesters, silicones, linseed oil, coal tars, and polyurethanes.[38] Some systems, such as phenolics, may be bake-cured to enhance the polymerization reaction. Resins such as epoxies, polyesters, and polyurethanes may also be cured by action of a catalytic hardener. Infiltration of the sealer into deep capillaries may be aided by vacuum impregnation, in which a vacuum is pulled on a component immersed in the sealer to remove air from the pores. Sealing of zinc and aluminum is made more effective by the use of a wash primer such as phosphoric acid,[38] acid zinc-chromate[112] or Steel Structures Painting Council (SSPC)[(4)] Paint 27 vinyl butyral.[114] Anaerobic methacrylate sealers such as Loctite[†] 290 have reportedly been highly successful in some applications.[115] These products have low viscosity in contact with air but harden when deprived of oxygen in coating pores.[38] The Loctite product has long-term thermal stability up to about 450°F (232°C).[115] Silicone-modified furan has been very successful in sealing coatings for digesters in Kraft paper mills, which are subject to caustic

[(4)] Steel Structures Painting Council, Pittsburgh, PA.

[†] Trade name.

stress corrosion cracking.[116] Waxes, greases, and oils may be appropriate for some machine part applications.

The most appropriate sealing strategy depends on service environment and compatibility with the coating. Most organics are limited to low temperatures. Aluminum pigmented silicone has been used to seal aluminum coatings exposed to temperatures as high as 535°C (995°F). For ceramic coatings, ceramic sealers such as sodium silicate,[117] heat cured organosilicates,[118] or aluminum pigmented silicate[119] may be appropriate.

Novel sealing strategies include laser glazing the surface of ceramics,[120, 121] sealing carbide cermet coatings with electroless nickel[122] or by infiltration of liquid solders,[38] and hot isostatic pressing (as is practiced to densify components made by powder metallurgy).[122] Mechanical treatments such as finish grinding[123] or peening[124] may be helpful in some cases.

Porosity of coatings may be used advantageously to retain release agents and lubricants such as PTFE, fluorocarbons, fluoropolymers, and silicone lubricants.[99]

Additional information on sealers is contained in the chapters on specific applications.

Mechanical Finishing

Surface finish is important for many coating applications involving wear. The smoothness attainable by finishing methods is dependent on process parameters used during spraying and on particle size range of feedstock (in the case of powder-fed processes). When surface contact with the mating part does not demand a high-quality finish, brush finishing may be an appropriate low-cost approach for softer coatings.[125]

Most metallic coatings can be finish machined. Cylindrical shapes are conveniently lathe turned, and other shapes can be dealt with if special care is taken at corners and edges.[38] Soft babbitts are machined using high-speed steel tools, while steels, brasses, bronzes and nickel-based alloys are finished more efficiently using sintered carbides.[126] Use of alumina tools loaded with titanium carbide is reportedly successful in overcoming machining problems for softer arc-sprayed steels.[127] Harder cobalt- and nickel-based alloys, including those containing tungsten carbide, can be machined with cubic boron nitride tools.[126]

High-quality finish, tight tolerance, and hard cermet and ceramic coatings may dictate grinding as the finishing operation. Diamond wheel grinding is recommended for the common ceramic coatings and various carbide cermets.[128] Careless finishing of ceramics can cause damage and result in inferior performance.[129]

A thorough treatment of recommended finishing practices is contained in the AWS Thermal Spray Handbook.[38]

Spray-and-Fuse

This hardfacing technique is the traditional domain of the welding community and is covered in depth in the AWS Thermal Spray Handbook.[38] Self-fluxing nickel-, iron- and cobalt-based compositions (some loaded with tungsten carbide) are called "self-fluxing" because their boron and silicon contents flux oxides on the splat and substrate surfaces. The fusing step, most commonly performed with an oxy-fuel torch, will densify the coating and enhance the bond to the substrate. Spray-and-fuse compositions generally have a wide melting range and are very viscous in the semimolten state. Proper control of heat input and timing during torch fusing is critical. Sagging and runoff, cracking due to thermal expansion incompatibility, shrinkage voids, distortion, excessive slagging, and base metal dissolution can all be problems. Furnace fusing and induction fusing are options for some applications and provide better control of the process. Induction is appealing because it heats a thin surface layer of the part.

The coating microstructure after fusing is essentially a cast structure and may have a materials response significantly different from that of the as-sprayed deposit.

Reactive Heat Treatment

In some cases, coatings may be heat treated to promote metallurgical bonding and formation of dense layers of new phases. For instance, aluminum coatings react with steel substrates to form iron-aluminum compound phase layers in a process akin to calorizing. The coating may be sealed with coal tar to prevent oxidation during the treatment.[38] In work with significant implications for the chemical process industries, Ishikawa et al. have explored various multilayer coatings in which dense diffusion layers formed during high-temperature service serve as effective barriers to the environment.[130]

CHAPTER 10

Safety

Safety considerations are covered in detail in the AWS Thermal Spraying Handbook.[38] Considerations include:

- gas handling;
- spray gun use and maintenance;
- abrasive systems hazards;
- personnel protection;
- toxic feeds.

Users should review materials safety data sheets provided with specific feedstocks and procedures manuals for any hardware prior to use.

One subject that merits further emphasis is the explosive hazards of fine metal dusts. Aluminum and zinc, in particular, are sprayed in large quantities and are readily combustible. Moist aluminum and zinc are capable of self-ignition.

Steffens and Hohle provide a specific example of an explosion incident in an operation for steel grit blasting and aluminum spraying of large oxygen lances.[131] Iron oxide residues reacted with fine aluminum by the "Thermit" reaction. In general, explosion hazards are avoided by adequate ventilation in spray booths and by wet collection using a water wash type of system. Good housekeeping should be maintained to avoid the accumulation of dusts.[38]

With increased attention to reactive metal compositions and controlled atmosphere, the potential hazards of explosive dusts need to receive increased attention. Inert shrouding and chamber spraying yield overspray with a more reactive and potentially pyrophoric surface character. Steffens and Hohle cite the example of vacuum chamber spraying of titanium.[131] A significant hazard exists when venting the chamber or opening the chamber door. Free dust accumulated on the chamber floor can be extremely dangerous. Fine dusts that enter the pumping system are combined with oil and rendered inactive.

Another source of concern is the possible sparking hazard associated with certain coatings, including aluminum, which could provide a source of ignition in potentially explosive environments. The pyrophoric character of titanium components under some conditions implies that this can be a problem for coatings as well. Engineers in the CPI should also be aware of the danger of having zinc in proximity to austenitic stainless steel. Fire or high-temperature excursions can trigger catastrophic liquid metal embrittlement.

CHAPTER 11

Coating Evaluation and Quality Control

Strict control of each step of the processing cycle is the key to assurance of coating quality and predictable performance in service. Simple lack of attention to the calibration of instrumentation is often responsible for problems.[132]

It is also important to recognize the value of property evaluations in determining if a coating meets requirements for acceptance in an application. As is usually the case for this technology, the stringent demands of the aerospace sector have prompted an ever-increasing level of sophistication. Applicators used to servicing this market are likely to follow the same procedures to ensure quality coatings for industrial applications.

Assessment of coating properties generally involves preparation of companion samples for destructive testing when the product coating is prepared. Some nondestructive methods may be applied to the coating itself. Most of these advanced techniques are just emerging from the lab stage, and it is unclear if they will have real near-term impact in any but the most high-tech applications.

Microstructural Assessments

A central theme of this book is that *coating performance is related to microstructure*. Inspection of prepared coating cross sections by optical microscopy or scanning electron microscopy is a means of detecting and measuring porosity, oxides, unmelted particles, and contaminations and delaminations at the coating/ substrate interface. While assessments are generally based on manual, qualitative evaluations, the most advanced metallographic systems include the capability of quantitative image analysis.[133] Unfortunately, the preparation of metallographic

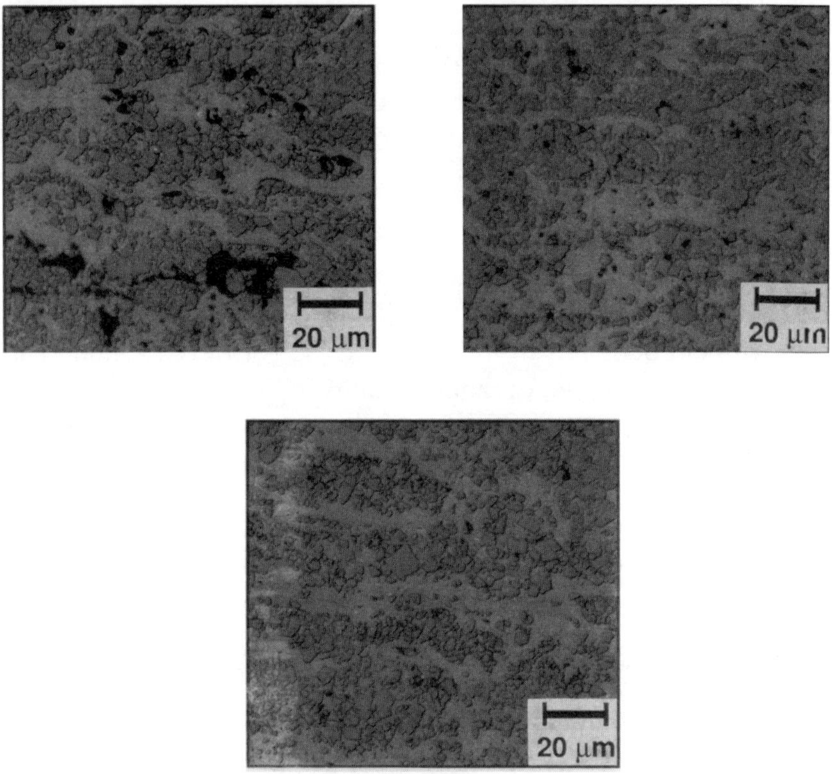

Figure 11.1 Comparison of WC/Co coatings prepared using three different grinding/polishing media: SiC paper (top left), bonded diamond platen (top right), diamond slurry lapping (bottom).[134] Used with permission of ASM International.

samples demands special attention if artificial defects—cracking, debonding, pullout, deformation, smearing, and embedded grit—are to be avoided. An example of the artifacts that can be generated is illustrated in Figure 11.1, which shows artificial porosity resulting from the pull-out of hard particles from a tungsten carbide/cobalt coating during polishing. The problem is reduced by using a bonded diamond platen rather than silicon carbide paper in metallographic preparation and is virtually eliminated by using diamond slurry lapping.[134]

Figure 11.2 further illustrates this effect for three different coatings, using data generated by quantitative image analysis. Figure 11.3 shows that the apparent microhardness of the tungsten carbide/cobalt coating is also a function of sample preparation.

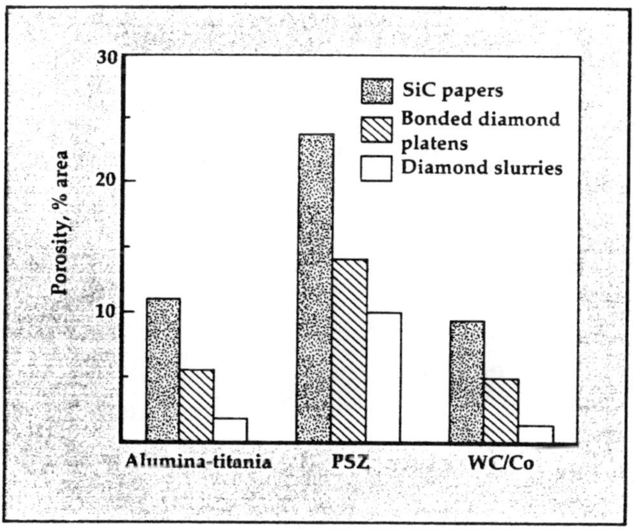

Figure 11.2 The grinding/polishing medium used for specimen preparation profoundly influences the apparent porosity determined by quantitative image analysis.[134] Used with permission of ASM International.

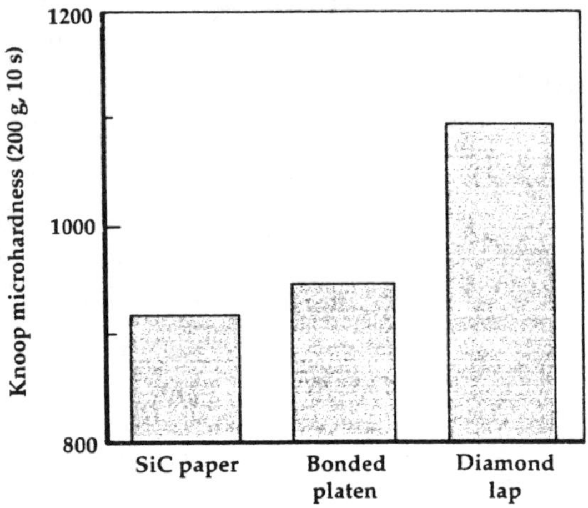

Figure 11.3 Comparison of microhardness results for WC/Co coatings. The hardness results for the specimens polished with diamond slurry are significantly higher than the measured results for the specimens prepared using SiC paper or bonded diamond platen.[134] Used with permission of ASM International.

Procedures for metallographic preparation of thermal spray coatings are currently being examined by the ASM International Thermal Spray Division responsible for this area.

Mechanical Properties

The standard tensile adhesion test (ASTM C 633)[(5)] is the most commonly employed method of quantifying coating mechanical response.[38] One surface of a cylindrical plug substrate is coated and the coating is bonded to a loading fixture with a suitable adhesive such as epoxy (Figure 11.4). The coating needs to be at least 0.015 in. (0.038 cm) thick to ensure that adhesive penetration into the coating does not extend to the substrate. Failure may be within the coating, at the coating/substrate inter-

(5) ASTM Standard C 633, "Test Method for Adhesion or Cohesive Strength of Flame-Sprayed Coatings" (Philadelphia, PA: ASTM).

face, in the adhesive itself, or a combination of these modes. The test is straightforward and allows coatings to be compared. Conoco has made successful use of a hydraulic adhesion tester to maintain quality control for critical tubular components of platforms in the North Sea and Gulf of Mexico.[135]

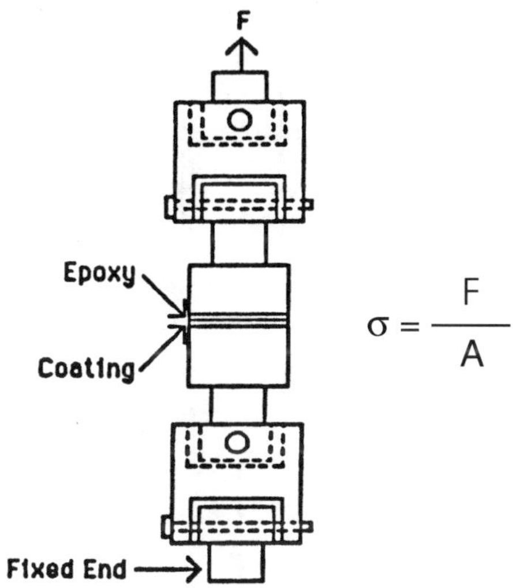

Figure 11.4 Tensile adhesion test.[136] Used with permission of ASM International.

However, the manner of loading in the standard adhesion test is not typical of coatings in actual service, and the method is subject to numerous sources of ambiguity and error.[38] For instance, results are dependent on alignment and degree of epoxy penetration. Proper preparation of the adhesive is critical. Berndt also points out that tensile adhesion is not a true materials property and that an insufficient number of data points is often used to draw statistically meaningful conclusions because of the typically large scatter.[137] Dissatisfaction with this method has prompted development of a number of alternative methods (Figure 11.5). For example, a shear test has been developed to determine fracture toughness under shear loads, which is believed to be a major cause of service failure. Some procedures are based on the fundamentals of fracture mechanics. While advanced techniques are largely intended to support research and development, it is possible that these will lead to standard practices useful for quality control and assurance.

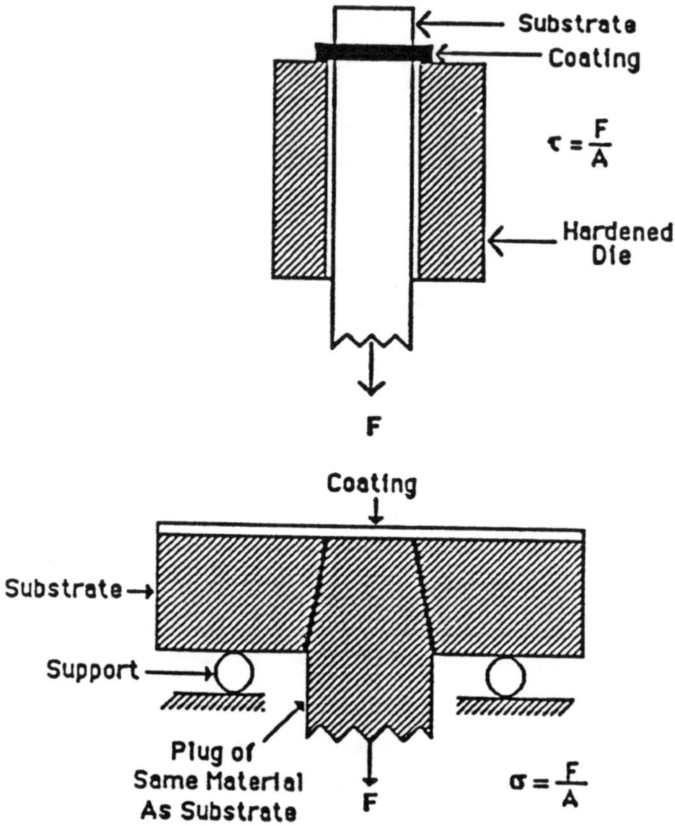

Figure 11.5 Nonconventional adhesion test configurations (above and opposite page).[136] Used with permission, ASM International.

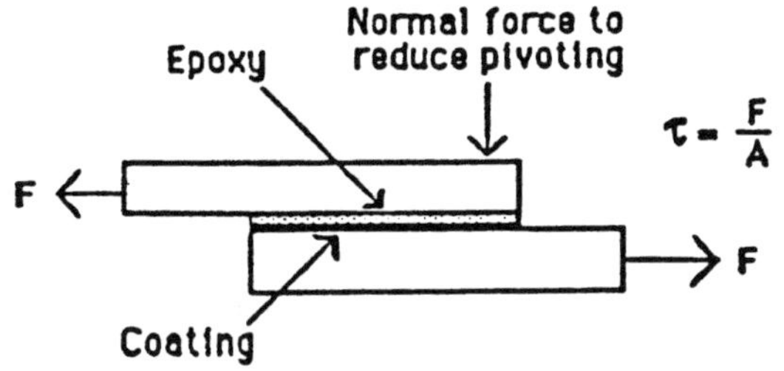

Lap Shear Adherence Test

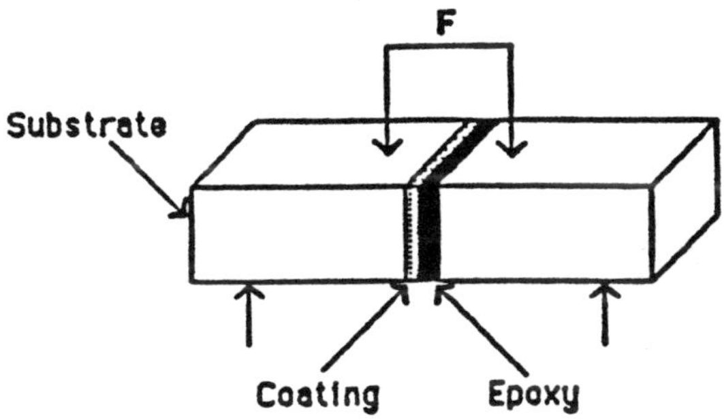

Four Point Flexure Test

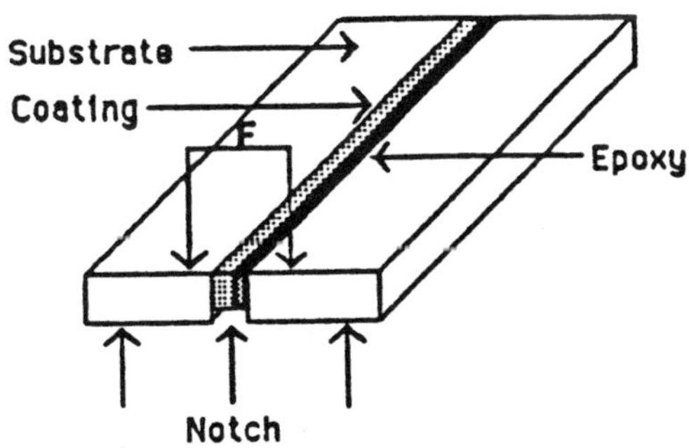

Double Torsion Specimen

Qualitative tests for coating mechanical integrity include chisel impacts, scratch tests, and crosshatching scratches on a surface and applying tape to lift any nonadherent coating. A useful test for aluminum or zinc on steel is to coat 0.050-in. (1.27-mm) steel coupons with 10 mil (0.25 mm) of coating and bend coupons 180 degrees around a $^1/_2$-in. (1.3-cm) diameter mandrel. The coating fails if it cracks significantly and can be lifted off with a knife blade. Strain tolerance of a ceramic coating may be assessed by a cupping test where the coating is indented with a 22-mm diameter ball.[138]

Macrohardness testing, usually by a Rockwell superficial hardness test, is a convenient means of gauging replication of coating characteristics but should not be looked on as a means of predicting wear performance or comparing different coatings. Microhardness indentation of specific features of a coating cross section is more likely to reveal meaningful information relative to performance. For instance, effects of local variability of quench rate can be disclosed in this way. Berndt et al. have noted that microhardness for a yttria stabilized zirconia coating is a function of location in the coating cross section, and that this measurement should be used with caution for quality assurance purposes.[139]

Accelerated thermal cycling tests may be employed using test coupons to give an indication of performance in thermal barrier applications.[138]

Nondestructive Testing Methods

As mentioned previously, many of the emerging nondestructive techniques for assessing coating quality are based on sophisticated methods, the results of which may be hard to interpret. Examples include ultrasonics,[134, 140] thermal imaging for thermal barrier coatings,[141-144] dielectric measurements for ceramics,[145] eddy current testing, and magnetic flux analysis.[38] A good example of a practical system is the prototype thermal wave scanning system for detecting nonbonded interfaces developed at the U.S. Puget Sound Naval Shipyard.[146] Forced hot air is used to inject heat, and a radiometer detects infrared radiation emitted by the coating. Disruption of heat flow at disbonds is displayed on a cathode ray tube (CRT). The method has been used for pumpshafts, valve stems, and plunges.

Assessment for Corrosion Applications

An area of acute interest to the CPI is protection of substrates in process solutions and other hostile corrosive environments. The first step in evaluating a new coating approach is generally a laboratory-scale corrosion test. Test coupons should reflect the coating variability expected in actual service, particularly with regard to porosity. In assessing a coated coupon's response, it is important to replicate the residual and service stresses encountered by the coating. The environment needs to be accurately simulated in terms of temperature excursions, fluid dynamics, concurrent wear, and solution chemistry. Trace impurities in real process environments (such as iron species in chloride process streams) can have a profound influence on materials response. Galvanic effects need to be addressed. The sacrificial effect of exposed substrate can cause misleading results when testing small coupons.

A laboratory test may also miss large-scale corrosion mechanisms such as concentration cells set up by variable access of oxygen.

In interpreting results of such laboratory corrosion tests, metallographic inspection should accompany physical examination and weight-change measurement. Attack of microstructural features such as intersplat regions, grain boundaries, and selected phases can compromise the protective value of a coating despite good apparent performance in a short-term test. The time dependence of corrosion response can be complex, and it is often improper to project a linear rate. For instance, depletion of a more active phase in the coating microstructure can completely change the mode of corrosion attack over time. In some cases, the corrosion product may be protective and give rise to parabolic kinetics, with the reaction rate decreasing as elapsed time increases.

For coatings that are not galvanically sacrificial with respect to the substrate, the presence of through porosity becomes a key issue. The technical literature contains some allusions to gas permeability as a means of detecting continuous paths from the exterior coating to the substrate.[118, 123] It is not clear from these references how the substrate is separated from the coating. In some cases it could be etched away, or the coating adherence could be minimized by using a polished substrate surface. Both these techniques could, however, alter the microstructural details of the coupon coating compared to the actual product coating.

Direct electrochemical measurements are another means of characterizing corrosion response. These should be capable of revealing galvanic coupling between coating and substrate and disclosing selective vs uniform attack of the coating microstructure. Electrochemical methodology is complex, and the details are beyond the scope of this work. A recent noteworthy study used polarization techniques to characterize the connected porosity of a ceramic coating.[147]

CHAPTER 12

Applications — Corrosive Process Environments

The primary limiting factor affecting thermal spray coating utilization in the chemical process industries is through-porosity associated with the as-sprayed deposit. This is especially critical when the coating is continually exposed to a corrosive electrolyte.

An effective sealing treatment is one approach to prevent through-porosity. Post-coating fusion, can in a sense, be regarded as a sealing process. Some coatings may offer a corrosion response that self-heals discontinuities. The porosity issue is moot if the coating is galvanically sacrificial with respect to the substrate.

However, galvanic coupling can cause other difficulties, such as evolution of hydrogen as part of the cathodic reaction. Galvanic effects can also have a dominant effect on performance if the anodic surface area is relatively small, as in the patching of a defect or the protection of weld areas of a vessel.

Elimination of through-porosity is the main target of advances in delivery technology such as vacuum chamber plasma-spraying and HVOF spraying with a controlled atmosphere shroud.

Even if porosity problems are negated, selective attack of coating intersplat zones, grain boundaries, or selected phases in the coating microstructure can still be a problem. The high quench rate associated with spray coatings can give rise to microstructures radically different from those of cast or wrought alloys of equivalent composition.

Coatings for General Corrosion Resistance

Harris of APS-Materials has provided overviews of performance of various air-plasma sprayed coatings in chemical process reactor vessels.[148] One example cited is a 3,000 liter carbon-steel vessel containing acetic acid and unspecified salts at 125°C (257°F) that was used in the manufacture of a commercial sealing product. An epoxy lining had failed in this application, while 0.5-mm plasma-sprayed "Hastelloy† C" gave four years of service. In a second application, plasma-sprayed Hastelloy B (UNS N10001) was used to solve a severe pitting problem at hot spots of an alloy 625 (N06625) vessel containing an unspecified acid at temperatures greater than 200°C (392°F). These examples are typical of the very cursory descriptions generally given in the published literature. It is likely that sealers were used in both cases, but these are not mentioned. One noteworthy detail from the second case history was that wrought alloy coupons with compositions equivalent to that of the coating had a corrosion rate of 0.175 mm/y, while the sprayed coating exhibited a rate of 0.0000075 mm/y! Significant differences in alloy microstructure between the wrought and rapidly quenched, as-sprayed materials may account for this dramatic change in response.

Nusum describes a successful application for wire flame-sprayed aluminum in the petrochemical industry.[149] Low-carbon steel valve shells and bodies of the Gutermuth design, subject to high concentrations of sour gases such as hydrogen sulfide, have a five-fold extension of service life by the deposition of a 20-mil (0.5-mm) coating. Again, no sealer was specified.

Glass and Depay of Dow Chemical have indicated the usefulness of the thermal sprayed ethylene acrylic acid copolymer in several process applications, including hydrogen peroxide and amine storage, latex process tank walls and blades, and petrochemical processing neutralization, fractionation acid, and drill-

† Trade name.

ing mud mix tanks.[105] It is also effective in protecting concrete exposed to sulfuric acid. The key to coating adherence and durability is control of heat input during spraying to avoid excessive crosslinking. The sprayed coating offers abrasion and impact resistance superior to epoxy polyamide and shows outstanding chemical resistance in a broad range of chemical environments. It is not recommended for use with chlorinated solvents, fuming or strongly oxidizing acids, aromatic alcohols, or heterocyclic aldehydes, and service temperature is limited to 165°F (74°C). Performance of the UTP Super Plast† nylon product has likewise been characterized in a broad range of chemical exposures.

In the pulp and paper industry, carbon-steel fiberboard press rolls may experience pitting corrosion. Wire flame-sprayed aluminum bronze plus an air-drying phenolic sealer, finished by machining, has given a 200 to 300 percent increase in service life.[125]

Smolak of Eutectic-Canada and Kammer of Eutectic Corp. recently described successful use of a novel nickel-based combustion spray coating (Metaceram Proxon† 21032) to solve a specific CPI corrosion problem.[150]

In processing of sour natural gas, sulfanol amine is used to remove hydrogen sulfide and carbon dioxide. Minor traces of acetic, formic, propionic, or lactic acid in sulfanol systems have produced corrosion/erosion rates of up to 0.2 in./y (5 mm/y) in low alloy steel vessels such as reboilers and stills. This problem has reportedly been overcome with a 20- to 25-mil (0.51- to 0.64-mm) coating and unspecified sealer. The coating is self-bonding and contains titanium dioxide, which may contribute to upgraded response to erosive wear. Deposition with the Eutectic Terrodyne† 3000 combustion spray system is claimed to offer advantages in terms of speed and ease of application.

MacMillan Bloedel has reported successful use of thermal sprayed nickel to protect carbon steel in "third effect" black liquor evaporators in a pulp mill.[151] Uncoated surfaces had been experiencing corrosion rates of up to 0.1 in./y (3 mm/y). A 15-mil (0.38-mm) coating, with 6 percent porosity and a 10-mil (0.25-mm) Plasite[†] 9570 sealer, has been in service for one year, and the only problem noted was the bubbling of sealer for one of the three evaporators under test.

Protection from Environmentally Assisted Cracking and Localized Attack

The pulp and paper industry provides an excellent example of the usefulness and limitations of thermal spray—in the prevention of stress corrosion cracking of carbon steel continuous digesters in Kraft paper mills.[116, 152-155] Cracking in this alkaline environment is largely associated with hard weld zones and attendant stress concentrations and residual stresses. In addition to type 309 (UNS S30900) and type 310 (S31000) stainless steel and Inconel[†] 72 weld overlays, thermal spray coatings have been successfully used to prevent stress corrosion cracking. Unsealed coatings quickly delaminated, but use of a silicone-modified furan sealer extended service life to more than three years.

McIntyre and Dillon also allude to the usefulness of flame-sprayed nickel in preventing stress corrosion cracking of carbon steel in hot caustic, although no specifics areprovided.[156] Presumably, a sealer such as furan was utilized.

Harris describes cooperative research of APS and Conoco to develop coatings for 1025 (UNS G10250) steel sucker rods for sour wells in East Texas.[148] In a laboratory evaluation, type 316L stainless steel was applied to tensile specimens by high-energy, high-velocity plasma-spray and sealed with furan. The test cycle involved stressing to 75 percent of ultimate tensile strength in 5%

NaCl saturated with respect to nitrogen and carbon dioxide. The uncoated 1025 standard lasted 9×10^5 cycles. The wrought type 316L stainless steel standard performed slightly better but failed by stress corrosion cracking due to chlorides. The coated test pieces survived past 4×10^7 cycles, the expected life of the wells.

Van Droffelaar and Hudson of DuPont Canada published extensive work on sacrificial coatings for type 304 stainless steel heat exchanger tubes in chloride solutions under acid or near-neutral conditions.[157] Best results were obtained with a Pb-2% Zn-2% Sb coating applied by thermal spraying or hot-dipping. In one case, service life was extended from 6 to 8 months to more than 10 years. This work dates from the 1960's, and spraying with a lead-based alloy is now less viable from a health/environment perspective. Also, the presence of a low-melting alloy containing zinc in association with austenitic stainless steel is a cause for concern with respect to possible catastrophic liquid metal embrittlement. However, the zinc in this case would be present as relatively stable zinc-antimony intermetallic compounds.

Tani et al. described another approach to mitigation of stress corrosion cracking of austenitic stainless steel.[158] Type 304 stainless steel was arc-sprayed with aluminum and then heat treated to convert the coating to iron-aluminum intermetallics. A typical heat treatment regime was 1 hour at 850°C (1,560°F). Slow strain rate tensile testing in room-temperature 40% $CaCl_2$ and U-bend testing in boiling $MgCl_2$ showed a dramatic improvement in stress corrosion cracking response for the heat-treated coating relative to uncoated and as-sprayed samples. In the U-bend test, a 100-fold improvement in time to failure over the uncoated stainless steel was reported.

McGuire alludes to thermal-sprayed aluminum sealed with silicone as useful for preventing external stress corrosion cracking of austenitic stainless steel under insulation at temperatures under 900°F (480°C).[159] This is an alternative to the technique of wrapping with aluminum foil.

Experience with Shrouded HVOF Systems

In the early stage of its Ultracoat†-shrouded HVOF development, Amoco performed extensive laboratory corrosion testing of type 316L stainless steel and "Hastelloy† C" coatings.[68] Sample configuration involved a carbon steel plate perimeter welded to a stainless steel backer plate. Test environments are summarized as follows:

Table 12.1

Solution	Concentration	Temperature	
	Vol. %	°F	°C
Acetic Acid	20	70	21
Acetic Acid	85	425	218
Acetic Acid	99	70	21
Sulfuric Acid	2	210	99
Hydrochloric Acid	1	95	34
Nitric Acid	7	150	66
Hydrofluoric Acid	0.5	200	93
Carbonic Acid	(pH=4; CO_2 purge in dist. water)	200	93
Sodium Hydroxide	5	200	93
Sulfur/Water Slurry	-	70	21
Molten Sulfur	-	300	149

Wrought type 316L, Hastelloy† C-276 (UNS N10276), and carbon steel were used as comparative standards. Exposure time was 336 hours.

Ultracoat 316L coatings were equivalent to or only slightly inferior to wrought 316L in acetic, carbonic, and nitric acids and in sodium hydroxide. High corrosion rates of the sprayed 316L in sulfuric, hydrochloric, and hydrofluoric acid were due to severe intersplat attack, which Amoco researchers equated with the presence of ferritic phase at splat boundaries.

Ultracoat "Hastelloy C" performed like wrought alloy C-276 in all solutions except nitric acid. Some cracking of coated weld regions was noted for sprayed "C" and stainless steel coatings in the 7% nitric acid and 85% acetic acid runs. This may have been triggered by local stresses caused by differential thermal expansion of the test sample. In liquid sulfur, Ultracoat 316L corroded at 2.6 mpy (0.066 mm/y), while Ultracoat "C" exhibited a rate of 1.4 mpy (0.036 mm/y). In the sulfur/water slurry, both coatings corroded at less than 0.1 mpy (0.003 mm/y).

Other lab tests at Amoco included thermal cycling of type 316L stainless steel coatings on carbon steel with very encouraging results. In one test, a specimen was heated to 600°F (316°C) and quenched in distilled water each hour over the course of an eight-hour run. No cracking was exhibited. In tests conforming to NACE standard TM0284-87,[5] Ultracoat deposits were found to significantly reduce hydrogen blistering of steels exposed to hydrogen sulfide.

Another lab test involved repair of stress corrosion cracked 300 series stainless steel tubes by coating with alloy C. Thermal cycling by internal heating of tubes to 90°C (194°F) in brine or distilled water showed effective bridging of the pre-existing cracks.

In another test simulating a gas well environment, N-80 steel coupons half-coated with "C" or type 316L stainless steel were exposed in a test chamber held at 250°F (121°C) and 430 psi (2.96 MPa) with an atmosphere of 5%CO_2 + 95%CH_4 over a 3% NaCl solution. There was no evidence of accelerated attack of the exposed carbon steel with the "C" coating present, while the

[5] NACE Standard TM0284-87, "Evaluating of Pipeline Steels for Resistance to Stepwise Cracking" (Houston, TX: NACE, 1987).

stainless steel coating triggered increased corrosion of the exposed area of the substrate.

General success of the lab-scale work triggered additional coupon evaluation and repair patch testing in a range of production environments. Ultracoat deposits have given excellent performance in the following applications:

Table 12.2

Ultracoat Tests in Production Environments

Unit/Part	Coating	Test/Repair	Duration
Solvent extraction unit	316L	Probe coupons	3 months
Hydrotreater - process side	316L	Two 1-ft^2 (0.09-m^2) patches on high pressure separator	2 years
Resid hydrotreater	316L	Test coupons	1 years
Crude tower	C	Test coupons	14 months
Hydrotreater - MEA service	C	Test coupons	3 months
Reformer - Flare knockout drum	C	Patches in drain area	In service 2 yrs
Asphalt unit-flare - Flare knockout drum	C	Stressed & unstressed hard weld samples (no HIC[(A)])	14 months
Liquid sulfur rail car	316L & C	six 1.5-ft.^{2} (0.14-m^2) test patches	19 to 21months
Sulfur plant Condenser head	316L	25-ft^2 (2.3-m^2) repair of 7-ft (2-m) diameter Weld HAZ[(B)]) & impingement area	21 months
Reformer	C	Two motor-operated valve stem shafts	4 years

[(A)] HIC = hydrogen-induced cracking

[(B)] HAZ = heat-affected zone

Negative results were encountered in applications involving a terephthalic acid unit (coating disbonded) and dimethyl batch reactor (coating attacked). Tests are reportedly ongoing in a number of other applications, including a maleic anhydride reactor (type 316L), weld protection of a sulfur recovery unit in MEA service (amine reboiler and regenerator, type 316L), and various components in an alkylation unit ("C").

CHAPTER 13

Applications - Wear

Mechanical components do not simply "wear out," they deteriorate by a variety of wear mechanisms that the relatively new technical discipline of tribology is seeking to elucidate. Alleviating wear is a primary means of enhancing durability and reliability of mechanical systems, reducing plant maintenance costs and down-time, and increasing industrial competitiveness. Development of materials more resistant to wear is a major objective of current research and development efforts relating to materials science and engineering.

The common perception that wear resistance is directly dependent on surface hardness is a gross oversimplification. Recent materials research has disclosed the complex relationships between a material's micro-scale constitution and a variety of possible wear mechanisms. These mechanisms fall into four general categories:[160]

Abrasion: produced by hard particles or protuberances forced against and moving along a solid surface.

Erosion: involving mechanical interaction with a fluid stream.

Adhesive Wear: initiated by localized bonding between surfaces in relative motion. Often this gives rise to debris that accelerates surface deterioration via "galling."

Surface Fatigue: associated with local fracturing caused by cyclic stresses caused by repeated rolling or sliding on a surface. Fretting, a form of sliding wear involving small relative motion of surfaces, may involve material removal by fatigue or corrosion-assisted breakdown.[161]

Wear performance is related to the nature of the mating surface, to the surface finish or roughness, and to the presence and continuity of lubricating films. Temperature, corrosion reactions, and atmospheric interactions (oxidation, sulfidation) are also important. Case histories involving combined wear and corrosion are given in Chapter 14.

Ceramic coatings deposited by thermal spraying often have outstanding wear characteristics and chemical stability. However, the tendency of ceramics to fracture under tension limits the range of application. Relatively soft coatings like aluminum bronze are appropriate for some applications. Analogous to hardfacing formulations, the key wear-resistant thermal spray coatings include the Stellite†-type cobalt-based alloys, the Tribaloy† cobalt- and nickel-based systems with Laves' phase strengthening, and composite "cermets" with tungsten carbide or other carbides in a tough metal matrix such as cobalt or nickel-chromium. Molybdenum-based materials are used for such applications as piston rings because of good resistance to adhesive wear and low coefficient of friction against steel and cast iron.[41] When the substrate can tolerate the temperature of coating fusion, self-fluxing spray-and-fuse alloy formulations may be used. These are generally nickel-based and may contain hard particles such as tungsten carbide. A great deal of the recent interest in wear-resistant thermal spray coatings has been prompted by concern with environmental problems associated with hard chrome electroplating.

Wear Mechanisms – Key Technical Considerations

Crook points out the importance of distinguishing between "low-stress" abrasion (during which the abrasive particles remain

† Trade name.

largely intact) and "high-stress" abrasion (in which abrasive particles fracture and constantly create new sharply angular morphologies).[161] Many of the common heat-treated steels, austenitic and duplex stainless steels, and single-phase nickel-based alloys perform poorly under low-stress conditions. Coatings containing hard particles, such as carbides and borides, do well under these conditions. However, the benefits of such microstructures are diminished or lost under high-stress abrasion.

On the other hand, some of the common alloys used in the CPI rival the performance of tool steels under these conditions. For pure metals and alloys, surface hardness can be a useful criterion for judging resistance to abrasive wear regardless of the stress mode. A surface hardness to abrasive hardness ratio of 0.6 or greater correlates with high wear resistance. However, the relationship of surface hardness to performance is not straightforward for multiphase materials, especially heat-treated steels. Crook indicates that even for single-phase materials, maximum hardness in the cold-worked state may be more relevant than actual hardness prior to service because of the work hardening involved in the abrasive wear process.[161]

Under the category of erosive wear, cavitation erosion has received a good deal of recent attention. The turbulent flow induced by pumps and turbines can induce local pressure drops below the vapor pressure of the liquid, nucleating gas bubbles. The shock waves from the collapse of these bubbles causes the surface damage. Berndt has provided an overview of the issues involved in selecting thermal spray coatings for this application.[12] After a break-in period, ceramics generally perform better than metallic coatings. The main problem for metallic coatings is the presence of oxide films between splats, which provides preferred paths for crack propagation. Spray-and-fuse nickel-chromium was found to be the best of the metallic materials tested. For plasma-sprayed oxides, resistance to nucleation and growth of intercrystalline cracks is the key consideration. Aluminum oxide, which has high inherent strength, and alumina-zirconia, which has a dispersion of submicron zirconium oxide, are good performers. Zirconia, with a relatively low 2% loading of yttria, also gives excellent performance. This may relate to the presence of some metastable tetragonal phase which transforms to monoclinic phase under the action of the cavitation stress. For this application, the fact that ceramic splats are incompletely bonded to each other may be an advantage that allows some elastic response under the oscillating stress field.

Wear under the condition of relative sliding of mating surfaces can involve both adhesive and fatigue-associated surface damage mechanisms. For metallic coatings, oxide development is important when surface temperature is high because of environment or frictional heating. If conditions are such that the surface oxide film remains intact, very little wear debris may be generated. On the other hand, stresses may be high enough to strip away the oxide, which rapidly reforms, leading to heavy wear and generation of oxide powder debris.[161] The porous nature of the thermal spray coating provides a mechanism for loading with release agents and lubricants such as silicones and curable PTFE and other fluoropolymers.[99] These may assist in upgrading wear response.

Case Histories

Cermet coatings containing tungsten carbide have a considerable track record in industrial applications, especially for protection of pump components. These materials are generally used below 1,000°F (540°C), with chromium carbide used above this temperature.[125] Ritchie of Bender Machine gives brief examples of HVOF-applied WC/Co for pumpshafts, impellers, compressor cylinders, fan housings, gripper fingers, and steam turbine runners.[162] A number of recent technical articles also expressed enthusiasm regarding HVOF-sprayed WC/Co.[163-165] HVOF is appealing for on-site premium wear coatings because it requires less sophistication and shows less technique dependence than other application methods.[47] It offers coatings superior to conventional plasma and comparable to those obtained via D-Gun.[166] Stellite product literature lists several oil patch applications for WC/Co applied by the Jet Kote[†] II HVOF system, including pump rods, impellers, seals, plungers, diamond drills, and reciprocating rods.[167] Another successful example of HVOF-applied WC/Co is in the protection of a draft fan from severe abrasion in a bitumen/sand processing operation.[149] It is important to keep in mind that the performance of these tungsten carbide cermet coatings is sensitive to both application technique and specific type of powder. For plasma-sprayed coatings, recent lab work indicates that a fine agglomerated powder performs best in the abrasive mode.[81] Under erosive conditions, cast and crushed material is undesirable because of the higher porosity of the coating.

Tucker of Union Carbide elaborates on pump component applications.[168] Tungsten carbide cermets and chromium oxide mate well with most packing materials used in centrifugal pumps, resulting in little adhesive wear. Sewage pump shaft sleeves of D-Gun coated WC/Co sealed with epoxy on type 400 series stainless steel have given 15 years of service. The same coating system has been used effectively for type 316 (UNS S31600) stainless steel wear rings, casing vanes, and packing box surfaces in a pump for recirculating catalyst-loaded slurry in an oil refinery, giving a three- to six-fold improvement in service life. Plasma-sprayed chromia has been used to combat adhesive wear in the seal area of an impeller for a seawater pump.

The Corrosion and Materials Engineering Committee of the Technical Association of the Pulp and Paper Industry's (TAPPI)[(6)]

[(6)] TAPPI, Atlanta, GA.

Corrosion and Materials Engineering Committee has documented case histories of thermal spray coatings in the pulp and paper industry,[125] some of which are summarized as follows:

Vibrating Chip Bin Bottoms: Plasma-sprayed alumina-titania on mild steel prevents roughening of the bin bottom which would cause plugging. The coating is disc-sanded. Five years of service without problems is reported.

Chipper Discs: A self-fluxing alloy with tungsten carbide applied by flame spray-and-fuse increases life of hardened alloy steel by 400 to 500 percent.

Chipper Blades: A tungsten carbide cermet applied by plasma and finished by diamond grinding has been used to upgrade response to abrasion.

Defibrator Digester Rotors and Agitator Digester Shafts: These stainless steel components are coated by the powder spray-and-fuse technique, presumably using a Ni-based self-fluxing powder.

Stainless Steel Rotary Valves in Pulp Mill: A 700 to 800 percent increase in service life is attained by using plasma-sprayed alumina with a nickel-chromium bond coat.

Slitter and Core Winder Mandrels: Powder flame spray-and-fuse coatings are used to combat abrasive wear.

More recently, Dawson and Christian reviewed hardfacing in the pulp and paper industry, including thermal spray techniques.[169] Unfortunately, details of these applications are sketchy. Some examples of hardfacing of rolls that involve thermal spraying include:

Paper/Fiberboard Rolls: Low-carbon, Co-based plasma transferred arc;

Anilox Rolls: Plasma-sprayed chromium oxide;

Corrugating Rolls: WC/Co cermet applied by plasma or D-Gun;

Grooved Press Rolls: Arc-wire sprayed austenitic stainless steel;

Press Rolls: Combustion-wire sprayed (probably austenitic stainless steel) phenolic sealer.

These applications involve high-speed, high-load abrasion from fillers, fibers, and doctoring. Corrosion and heat transfer may also be at issue.

Recently, Scholl and Clayton reported experimental work aimed at improving the efficiency of the pulping process by changing the texture of refiner plates via high-energy plasma-spraying.[170] By manipulating powder particle size and spray distance, surface roughness was varied for several different metal, oxide, and cermet coatings. There was no correlation between surface roughness and abrasive or erosive wear. It is not clear whether pulping efficiency was improved by the use of rougher coatings.

D'Angelo and El Joundi[99] and Hamel[171] provide good overviews of oxide coatings for wear resistance. Chromium oxide, applied by plasma-spraying or the Rokide rod-feed combustion system, is the workhorse coating composition. Its high hardness corresponds to outstanding resistance to abrasion. Its coefficient of friction is low, and chromia has outstanding self-mating and antigalling properties. Rokide product literature lists numerous applications in the petrochemical and pulp and paper industries, mostly involving pump components. For example, chromia is a preferred alternative to chrome plating for protecting pump rotors and shafts used for delivery of starch and clay fillers for pulp stock. Bronze pump sleeves badly grooved by particulate in river water have been replaced by chromia-coated Mechanite,[†] with a dramatic improvement in performance. In the coating of paper with clay and titanium dioxide slurry, deterioration of the stainless steel end dams that engage rubber rolls has been minimized by application of chromia. Addition of silica to chromia provides a coating more durable with respect to impact. This can be important if components are subject to rough handling in an industrial setting. Titania is a tough, durable coating. It has been used for the bearing surfaces of grinder spindles.[99] Alumina has hardness and wear resistance nearly equivalent to that of chromia, plus good insulating properties. Hamel indicates that a bond coat such as nickel-chromium is generally used in industrial applications for oxide ceramic coatings, but this is not typically specified in references to case histories.[171]

A focus of recent interest has been alumina-titania. Titania additions to alumina decrease wear resistance but upgrade toughness. Coatings are very dense and produce a smooth surface as-sprayed. This minimizes the need for post-spray finishing. The coating finds use in protecting plungers for reciprocating and centrifugal pumps in petrochemical applications.

Heiman et al. report on process parameter optimization for plasma-sprayed alumina-titania.[172] Herman points out that a significant advantage for this system is its robustness with respect to application variables, making it easy to apply high-quality coatings.[129]

In an interesting recent study, Gansert and Herman compared HVOF and plasma-sprayed powder blends with 40% titania.[48] Analysis of the plasma-sprayed coatings indicated formation of aluminum-titanate, while the HVOF structure was essentially alumina embedded in a soft titania matrix. This composite structure corresponded to enhanced sliding wear resistance, as measured by a standard pin-on-disc test.

Several recent innovations in coatings for wear resistance are based on novel coating microstructures. Resistic† powders produced by Alloy Technology International consist of titanium carbide in a matrix of Ni-Cr, tool steel, or stainless steel. Pin-on-disc testing has disclosed superior sliding wear response compared to WC/Co and Stellite 6.[173] A number of novel iron-based coatings are emerging in the marketplace. Wear Management Services provides a Tribolite family of spray powders that are cobalt-free and have relatively low chromium. Sliding wear performance is markedly superior to Stellite 6 in HVOF and plasma-spraying.[174] As-sprayed coatings by plasma and HVOF are amorphous or microcrystalline. The mechanism for the superior wear response is unclear but may relate to a build-up of boron-rich phases in the wear track. By blending with Ni-Al, dense, well-bonded coatings can be attained by simple combustion torch. The real niche for this coating may be in severe, high-temperature wear applications associated with metalworking applications such as seamless tube manufacture and steel strip production. Arc-sprayed FeCrBSi cored wire[175] is similar to Tribolite.† As-sprayed coatings are partially amorphous or microcrystalline and show sliding wear response superior to Stellite 6. The Metamorphic† alloys marketed by Amorphous Metals Technologies are claimed to undergo a surface transformation from crystalline to amorphous in response to wear in service, with attendant superior response to sliding wear and abrasion by fine particulate and slurries.[86]

CHAPTER 14

Applications — Combined Wear and Corrosion

The task of designing a thermal spray coating to combat wear is much more formidable if service is in a severely corrosive environment. Unfortunately, this is often the case in CPI applications. The main problem is the through-porosity inherent in most as-sprayed deposits. If the coating is relied upon as a physical barrier to the corrosive medium, penetration to the substrate can lead to rapid degradation.

Several approaches concerning porosity can be considered:

Coating System: The premium approaches of D-Gun and chamber spraying may not be practical or cost effective for a given application. Some of the new advances based on HVOF technology may be useful, but this has not yet been extensively explored.

Post-Coating Treatment: Inadequate attention has been paid to sealing strategies. Heat treatments can generate fully dense barrier layers.

Self-Healing Coatings: Corrosion products of substrate and/or coating can plug pores.

Multilayer Coatings: The outer layer provides resistance to wear while the interior is relied upon for corrosion protection. Alternately, powder blends of ceramics and corrosion-resistant alloys may be sprayed to yield coatings resistant to penetrating attack.

Sacrificial Coatings: The coating protects exposed substrate by sacrificial action.

Coating Quality Control: A straightforward, nondestructive technique for detecting through-porosity would be highly desirable for upgrading reliability for these applications.

Some of these strategies are touched upon in the compilation of case histories. They are addressed again in Chapter 21.

Case Histories

Amoco's experience with ceramic wear-resistant coatings under corrosive conditions has not been overly positive.[176, 177] In an industrial setting, the relatively brittle coatings can be easily damaged by rough handling. Successful examples of applications include plasma-sprayed chromia and alumina on stainless steel in concentrated sulfuric acid. These coatings work well because the substrate will not be aggressively attacked if the coating is breached. Plasma-sprayed chromia has also been successful in protecting valve stems from attack by hydrochloric acid.

United States Navy in-service testing of plasma-sprayed shipboard steam valve stems has been documented recently.[178] The preferred coating system consisted of a prealloyed Ni-Cr-Al-Mo-Fe bond coat (Eutectic† 21031) and a ceramic topcoat of zirconia-titania-yttria (Metco† 143). Inspection after six years revealed superior performance compared to pitting of the uncoated type 410 (UNS S41000) series stainless steel stems after two years. The Navy's positive experience with plasma-spraying has also included protection of main feed pump shafts using a composite bond coat (Metco† 447) and alumina-titania topcoat (Metco† 130).

For pump shafts subject to hydrogen sulfide and hydrochloric acid in water droplets, Amoco found that a switch from WC/Co

† Trade name.

to WC/Ni-Cr cermet was beneficial. Lugscheider et al. provide additional examples of successful utilization of the Ni-Cr bonded cermet.[49] These include coatings for turbine pumps used in evaporation thickeners for waste sulfuric acid containing undissolved titanium sulfide, stirrers for nitric-phosphoric-fluorosilicic acid with calcium nitrate, and bucket wheel conveyers in alkaline solutions with dispersed abrasive particulate. Tokoi recommends a WC/Ni cermet for severe caustic, phosphate, and amine service where cobalt-bonded tungsten carbide does not hold up.[179]

Cummings and Brown have documented largely unsuccessful attempts to utilize plasma-sprayed and D-Gun coatings in seawater pumping applications.[180] Stainless steel pump bearing sleeves coated with chromia and run against silicon nitride pads were plagued by crevice corrosion of the substrate. Nickel-aluminum bronze sleeves coated with 8 mil (0.2 mm) of tungsten carbide that were run against silicon nitride blistered in a few hours. A stabilizer shaft coated with 6 mil (0.15mm) of tungsten carbide and an unspecified sealer that was run against a "ferrbestos" liner showed deep pitting after 18 months of service. Sealing under a vacuum was not beneficial. Attempts were made to use a corrosion-resistant base coat (Cu-30% Ni or pure Ni) for WC topcoats. Substrate corrosion could not be avoided. Based on this limited experience, a corrosion-resistant substrate like alloy 625 (UNS N06625) is required.

Recent United States Army work on coatings for lock and dam facilities points out the performance limitations of standard sealers for nonsacrificial coatings under abrasion plus immersion conditions.[116] Arc-sprayed aluminum bronze and type 300 series stainless steel coatings were evaluated in service with three different sealers: vinyl acetate-vinyl chloride copolymer; epoxy polyamide; and a three-coat system with SSPC Paint 27 vinyl butyral wash primer, vinyl intermediate coat, and V-766E topcoat. All showed pinpoint rusting and blisters or delamination after only nine months. Zinc and zinc-aluminum coatings sealed with the wash primer only show no rust after 10 months, but some polishing indicates that wear is occurring. A sacrificial coating with some enhanced resistance to abrasion would be desirable for this application.

Arsenault et al. at the Canadian Industrial Materials Research Institute have developed a Zn-30% Ni base coat for plasma-sprayed ceramics such as chromia and alumina.[90] The intermediate layer is galvanically sacrificial with respect to the carbon steel substrate, inhibiting the corrosion problems associated with coating porosity. The efficacy of this approach has been indicated by salt fog testing and electrochemical characterization, but no industrial experience has been reported.

Polish technologists have tackled the problem of combined corrosion and wear in a novel manner.[123] Standard oxide ceramic feedstocks were blended with corrosion-resistant metal alloys such as Ni-Cr, plasma-sprayed, and tested for 1,200 hours in various acids and bases typical of papermaking. The best performance was exhibited by chromia and an alumina plus 20% occluded nickel powder for H.C. Starck, each blended with 20% Ni-Cr. The chromia-based coating required finish grinding to render the coating completely resistant to nitric acid. This is an example of the practical innovation that is possible with a little creative thinking. The metallic islands in the coating may block interconnected porosity pathways to the substrate and also improve resistance to crack propagation. The Polish research is also noteworthy in pointing out the poor performance of coatings containing Ni-Al in acid environments. The various Ni-Al materials are the most commonly specified bond coats. If attention is not paid to the corrosion response of the bond coat, premature coating failure will often result, prejudicing the user against further application of thermal spray. Meringolo provides additional examples of successful use of Ni-Cr bondcoats with chromia topcoats in the pulp and paper industry.[125] These include components of an acid/lime mud centrifugal pump and a Nash rotary pump. M. L. Thorpe alludes to positive experience with 2 mil (0.05 mm) of flame-sprayed nickel-chromium under 12 mil (0.3 mm) of alumina-titania for valve faces subject to erosion by solids in pumped liquids in the petrochemical industry.[39] Unfortunately, the specific process streams are not identified.

P.H. Thorpe provides a good example of effective use of thermal spray in the Australian pulp and paper industry.[181] Continuous digesters for neutral sulfite "semi-chem" pulping are constructed of 3-mm type 316L (UNS S31600) stainless-clad carbon steel. Scraper action at the bottom head can lead to rapid removal of the protective cladding. Local plasma-spraying with Stellite[†] 1H avoids the problems with weld overlays and provides 6 years of service. This is an example in which a nonsprayed underlayer (cladding) provides the corrosion protection.

Another variant of this approach is described by McMordie of Sermatech International.[119] Aluminum oxide is sprayed on a proprietary, heat-cured aluminum loaded ceramic base coat which provides sacrificial protection for airfoils in a blast furnace air compressor subject to ammonia, sulfur compounds, coke dust, heat-treat scale, and rust. While this specific case history relates to a steel industry application, it could have use in some CPI applications as well.

Union Carbide product literature describes a number of wear plus corrosion applications for the premium D-Gun coatings in the petrochemical industry, mainly involving pumps and valves. For rotor shafts in centrifugal pumps and compressors subject to corrosive environments (SO_2, H_2S, Cl, etc.) and wet or dry particle contamination, an overlay composition of W-8% Co-3% Cr5% C (LW15) is recommended, with W-19% Cr-5% Ni-6% C used above 1,000°F (540°C). For gate valves (stems, gates, and seats) and ball valves (ball and seat) subject to H_2S, a key consideration is avoiding sulfide stress corrosion cracking by keeping substrate hardness low. D-Gun coatings based on chromia, tungsten carbide, and chromium carbide are utilized to provide the desired sealing, avoid wear degradation, and impart corrosion resistance. At high temperatures, the chromium carbide plus nickel-chromium composition is recommended.

Several recent innovations in feedstock are specifically targeted for service under corrosive wear conditions. Metco 700 and 700F amorphous, nickel-based plasma-spray powders are recommended for abrasive wear, scuffing, and aqueous corrosion, but they are not appropriate under erosive conditions. Stellite's Tristelle family of cobalt- and silicon-modified stainless steels (which form chromium-rich carbides) are suggested for various pump and compressor components, knife and scraper blades, and bearing applications.[46] Plasma Sulzer Technik has introduced a new MoFeCrNi prealloy powder (SP174, SP175) and relying on a novel two-phase microstructure for enhanced wear performance. Performance data is not yet available.[107] Anval has introduced Ultimet,[†] a cobalt-based alloy with high molybdenum. The composition has been optimized to provide both wear and corrosion resistance.

Greving and Shadley of the University of Tulsa recently documented an extensive laboratory evaluation of thermal spray coatings for oilfield equipment applications.[182] Erosion was characterized by sand/water slurry stream impact in accordance with ASTM Standard G 73,[(7)] while abrasion resistance was quantified by ASTM Standardized Test G 75.[(8)] This involves reciprocating motion of test panels over a surface covered with sand/water slurry. Corrosion testing was performed in an auto-

[(7)]ASTM Standard G 73, "Practice for Liquid Impingement Erosion Testing" (Philadelphia, PA: ASTM).
[(8)]ASTM STandard G 75, "Test Method for Slurry Abrasivity by Miller Number" (Philadelphia, PA: ASTM).

clave in a CO_2 and deaerated seawater environment and in aerated seawater. Seventeen common feedstocks were evaluated. Application techniques included flame spray-and-fuse, Rokide ceramic combustion process, air plasma-spray, and HVOF. Results are presented in Figures 14.1 through 14.3.

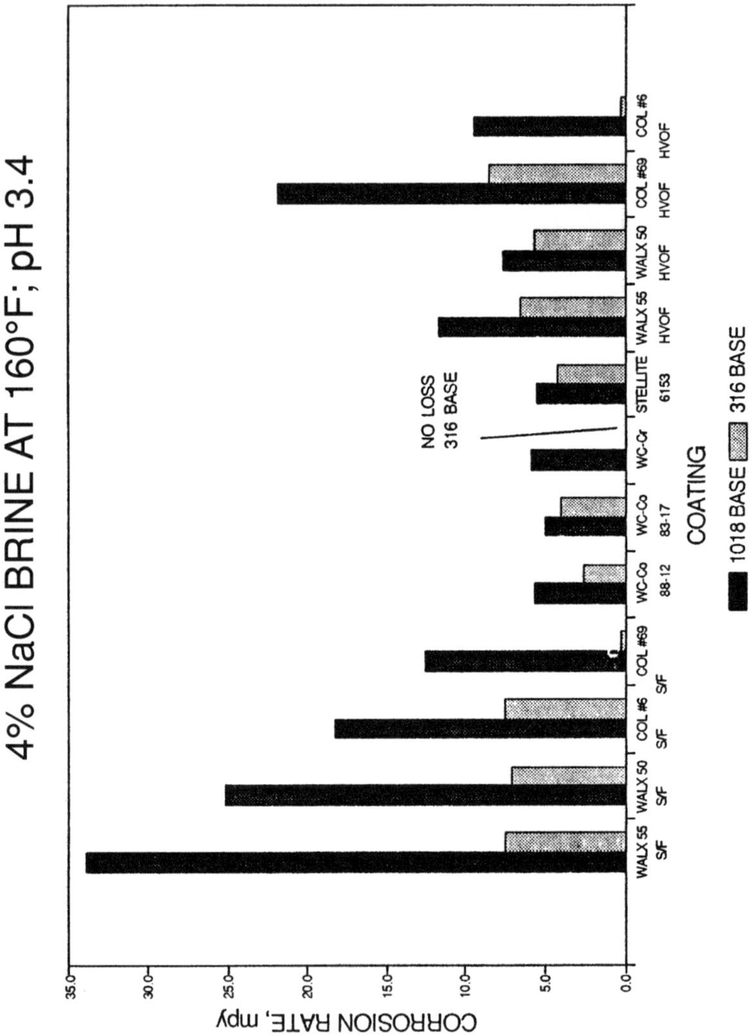

Figure 14.1 Corrosion test results; University of Tulsa evaluation of coatings for oilfield equipment. Courtesy of D.J. Greving and J.R. Shadley.

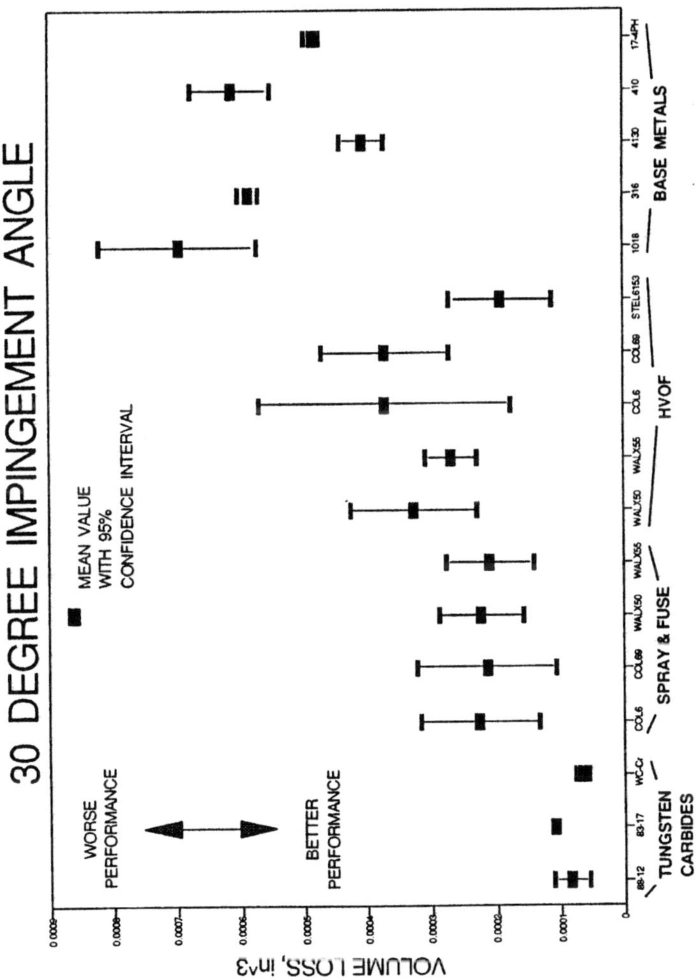

Figure 14.2 University of Tulsa erosion test results.

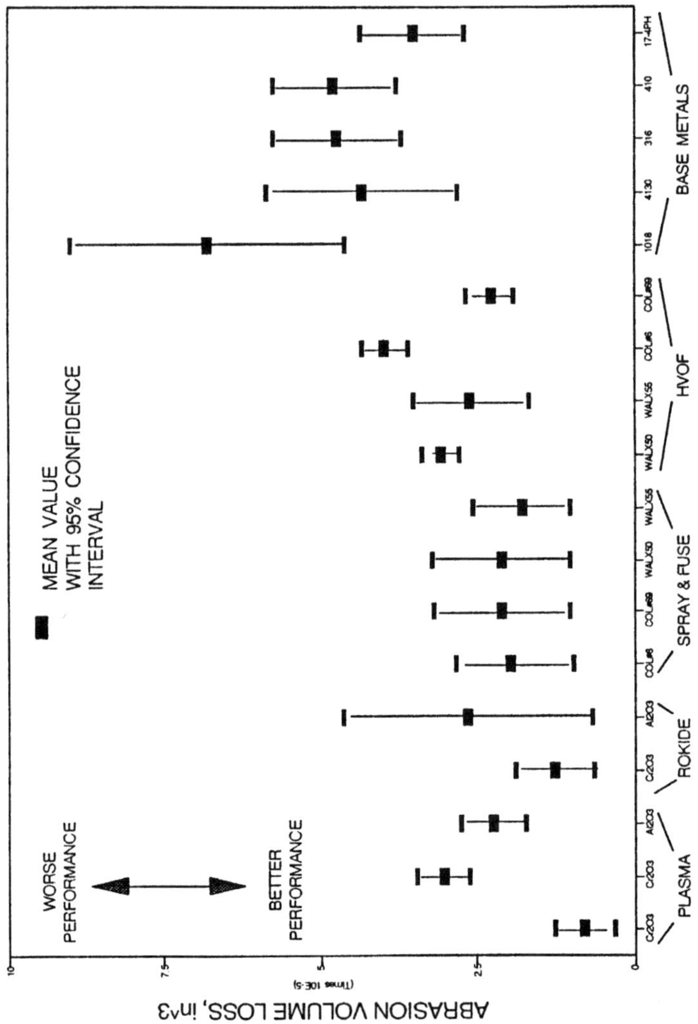

Figure 14.3 University of Tulsa abrasion test results.

CHAPTER 15

Applications – Restoration and Repair

Perhaps the earliest uses for thermal spray coatings were in the buildup of improperly designed or fabricated components and the restoration of worn machine parts to proper dimension. Significant advantages compared to weld repair processes include the absence of distortion and heat affected zones, minimal residual stresses and part oxidation, and advantageous overall economics. Required coating characteristics for these applications include low shrinkage, ability to attain thick deposits, good adhesive and cohesive strength, finishing characteristics compatible with the component, low-cost feedstock and application hardware, and high deposition rate and efficiency.

The real benefit of thermal spray is not fully realized, however, unless the coating is selected to upgrade wear response or otherwise generate a surface tailored to the specific application. For instance, Ritchie describes the common job shop restoration technique of building up worn hard chrome-plated shafts with type 420 (UNS S42000) stainless steel and then restoring the wear-resistant surface by replating.[162] A focus of recent development in thermal spraying has been the identification of coatings that can *replace* the environmentally noxious chrome plating process. In the pulp and paper industry, Yankee drier rolls have been resurfaced with arc-sprayed type 400 series stainless steels and finished by grinding. However, poor heat transfer leads to a 10 to 15 percent drop in production.[183] A new molybdenum-based coating applied by plasma-spraying of powder has overcome this heat transfer difficulty and also dramatically reduced the rate of wear.[183] The Yankee drier application is detailed in Chapter 18.

A focus of recent innovation in this category of application has been the use of composite self-bonding powders sprayed by combustion or plasma which yield thick, well-bonded, wear-resistant coatings. Examples include Metco[†] 447 (a molybdenum-nickel-aluminum composite powder) and 449 (a cast-iron core clad with aluminum, molybdenum, boron, and iron oxide). The newer Metco 449P product has a finer particle size to yield finer ground finish and a denser microstructure.

It is also possible to blend these powders with lower cost iron-based feedstocks to yield successful coatings. For instance, Wear Management Services markets a "one-step" variant of its Tribolite[†] wear-resistant iron-based powder. The blending with a bond-enhancing powder yields a feedstock that can be applied by very simple low-cost combustion torches. Coating quality is relatively insensitive to variation in spraying parameters. This innovation makes thermal spray restoration technology accessible to small job shops and plant maintenance departments. The success of this approach has been demonstrated in several industrial applications, including surfacing of steel mill strip conveying rolls.[7]

Arc-spraying is appealing for these restoration applications because of high deposition rate. Unger describes the usefulness of prealloyed Ni-5% Al wire feedstock for repair of aircraft engine components.[184]

In spraying of high-carbon steel feedstocks for repair applications such as crank shafts, a key consideration is the loss of carbon that occurs during the coating process. Wang et al. indicate that a loss due to oxidation of as much as 42 percent can occur.[29] They solved this problem in the case of arc-spraying by using nitrogen as the atomizing gas.

[†] Trade name.

Repair of damaged reaction vessels is a topic of special interest to the chemical process industries. Repair of glass-lined vessels is discussed in Chaper 19, and a remedial procedure for environment-associated cracks is described in Chapter 16.

CHAPTER 16

Applications – High-Temperature Environmental Degradation

Metallic coatings for high-temperature service generally rely on the presence of significant chromium or aluminum to generate self-healing protective oxide films.[185] In the aerospace industry, there is a wealth of experience in the protection of gas turbine blades from the oxidation and hot corrosion generally associated with sodium sulfate attack.[186] For this application, a preferred coating selection is "M-CrAlY" (with M = Ni, Fe, or Co), with vacuum plasma-spraying yielding the highest performance coatings.[185] In the CPI, there is a much broader range of service environments plus the need to protect large surfaces with reasonable economics. A primary focus has been on boiler applications, particularly the black liquor recovery boilers in the pulp and paper industry. High-chromium alloys have received a good deal of attention. Aluminum is often a cost-effective choice for high-temperature oxidation resistance. Oxide coatings have also been used, despite their generally poor fracture toughness. Refractory oxides are especially useful when high-temperature oxidation is coupled with erosion or other wear-related degradation. Oxide coatings are, however, prone to stress problems relating to thermal shock and thermal cycling. Another problem with which coatings can contend is environmentally triggered cracking, which is often a problem in reaction vessels.

Utility Boilers

Experience with thermal spray in the utility boiler applications has some relevance to the CPI. ASB Industries of Ohio has reported on HVOF spraying of boiler superheater tubes with types 316 (UNS S31600) and 420 (S42000) stainless steels.[187] Extensive research on plasma-sprayed coatings for coal-fired boilers was conducted in the early 1980s by the British Central Electricity Generating Board in association with Plasma-Technik.[188] Based on experience with wrought alloys, high-chromium content alloys were evaluated for furnace walls and superheater tubes exposed to reducing and sulphidizing atmospheres. Best results were obtained for a 50% Ni-50% Cr alloy, and repair of deteriorated tube surfaces was found to be cost effective compared to tube replacement. Documentation of this work emphasizes the importance of careful spray process control to ensure reproducible high-quality coatings.

A related development by this same group is a hybrid combination of plasma-spray and plasma transferred arc, in which the transferred arc plays a secondary role in developing a dense, well-bonded coating.[189] A self-fluxing nickel-based alloy with 15% to 18% chromium did not show the layering, oxides, and porosity typical of as-cast or welded overlayers and had a much thinner heat-affected zone than conventional plasma transferred arc. Coating thickness was limited to 750 µm without spalling, but this approach was felt to be cost effective compared to clad material in less severe applications.

In what appears to be a refinement of the above research, Japanese workers have described a multilayer coating for surfaces subject to hot corrosion caused by low-grade fuel.[190] The bond layer is Ni-Cr, with a second layer being a blend of Ni-Cr with 30% Al_2O_3. An optional third layer of straight aluminum oxide is also described, presumably for enhanced erosive wear resistance.

Black Liquor Recovery Boilers

One specific industrial application for which thermal spray approaches have been aggressively pursued is the protection of lower furnace sections of pulp mill black liquor recovery boilers. This is an extremely severe high-temperature service environment, with incomplete combustion of high-sulfur fuel, fluctuation between oxidizing and reducing conditions, and presence of a highly corrosive, fused "smelt" containing sodium sulfide.[191, 192] The cost of repair and down time for this critical plant subsystem have motivated a continuous evolution of coatings for protection of waterwall tubes and other furnace surfaces since the 1960s. Initial experience with nickel-based plasma-sprayed coatings such as Metco† 444 (9% Cr, 5.5% Mo, 7% Al, 5% Fe), Metco 47C (16.5% Cr, 17% Mo, 6% Fe, 4.5% W), and Combustion Engineering 2148 (16% Cr, 16% Mo, 4% Al, 6% Fe, 4% W) indicated their unacceptability for this application. The early technical literature also contains an intriguing reference to promising field testing by Combustion Engineering of a plasma-sprayed pure chromium top coat over a "nickel-titanide" bond coat.[94] Another multilayer approach that gained significant commercial acceptance was the "Multicoat" system, consisting of flame-sprayed layers of Ni-20% Al bond coat, type 300 series stainless steel, and a pure aluminum topcoat (often sealed with aluminum powder in a silicone or silicate carrier).[94, 191]

Industrial experience with the Multicoat system was not completely satisfactory in terms of coating life.[191] The coating reportedly gave four years of service but required annual repair.[94] The main problem was apparently corrosion-induced spalling caused by subsurface corrosion related to coating through-porosity. In response to this continued market need, TAFA embarked on a development program in association with International Nickel and Metalspray that focused on arc-spraying of high-chromium, nickel-based wire. As cited, work by the British Central Electric Generating Board indicated that a chromium content in excess of 40 percent is required to protect nickel-based boiler alloys and coatings in the presence of high-sulfur fuel.[188] The fruit of the TAFA development program is the nickel-based

† Trade name.

wire feed "45CT," with 45% Cr and 1% Ti for bond enhancement. TAFA literature promotes the advantages of arc-spraying for this application in terms of bond strength, coating porosity and oxide inclusions, and forgiveness with respect to the less-than-ideal conditions when spraying inside a boiler.[191] TAFA alludes to a five-year warranty for this coating.[19]

Another candidate for this application is the iron-based Kanthal† M wire, with 22% Cr and 6% Al for bonding.[192] Field application has taken the novel approach of dual spraying with electric arc for enhanced bonding and a flame-sprayed top layer, which is claimed to give better coating quality in terms of porosity and oxide content. Published work cites positive results of industrial trials after nine months, but it is unclear if longer-term performance was acceptable.

Metco offered a similar composition (also containing 2% Mo) as a powder for plasma-spraying and also proposed composite powders of Fe-Cr clad with Al-Mo or Al-yttria for dense, well-bonded coatings.[193, 194] Metco publications do not cite actual industrial trials for these powders.

Based on the continued effort in coating development for this application, it appears that wire-arc applied TAFA 45CT does not provide a final answer. As part of its Ultracoat† development, Amoco workers addressed two key issues affecting coating quality. Coating quality is optimal when spraying flat surfaces, but porosity tends to increase when the angle of incidence is not close to normal, as in the "crotch" areas of waterwall tubes. Amoco demonstrated that the inert gas shrouded Ultracoat HVOF delivery system is much more forgiving than other spray hardware with respect to spray angle, and provides a generally superior coating in terms of porosity.

The second issue is the general difficulty in maintaining parameters such as stand-off distance when spraying inside the boiler. Amoco developed a robotic system to control the gun position with respect to the wall configuration.[68]

Another recent commercial development for on-site spraying is the Metalspray Densys† system, which is believed to be an HVOF-based technology reminiscent of Ultracoat. No details are available.

Aluminum

For general high-temperature protection of steel, the formation of a dense, self-healing aluminum oxide film makes aluminum the coating of choice. Service environments up to about 900°C (1,650°F) have been indicated in the technical literature, with service in sulphur-bearing environments limited to 800°C (1,470°F).[195] The high-temperature aluminum-pigmented sealer used in conjunction with sprayed aluminum is limited to 480 to 535°C (896 to 995°F).[195, 196] At higher temperatures, a diffusion heat treatment to convert the coating to dense aluminum-iron compounds is recommended.

Literature from Hobart-TAFA indicates that the superior interface bond of arc-sprayed aluminum allows this diffusion process to occur in service rather than in a separate heat-treating step. The reaction between aluminum and the steel substrate to form intermetallic compounds is illustrated in Figure 16.1.

For applications above the melting point of aluminum (660°C, 1,220°F), attention clearly needs to be paid to the rate of heating so that the conversion to aluminum-iron compounds occurs properly and melting of the coating does not cause problems. In the range of 900 to 1,095°C (1,652 to 2,003°F), TAFA literature recommends a nickel-chromium coating followed by aluminum or FeCrAlY in the presence of sulphurous gases.[196]

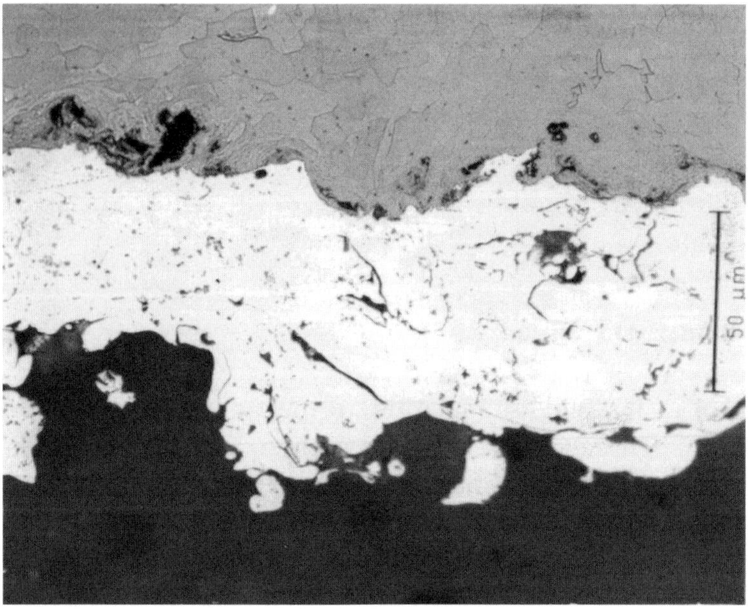

Figure 16.1a Thermal conversion of arc-sprayed aluminum coating as sprayed. Note the partial disbondment at the steel/coating interface and the formation of intermetallic compounds. 500X. Courtesy of Intermet Technology.

Other Multilayer Approaches

Several coating strategies involving multiple thermal spray layers and/or innovative sealers have already been described. Arcolin et al. reported on the effectiveness of plasma-sprayed chromium with a novel ceramic paint sealer for protection of type 304 (UNS S30400) stainless steel coils in ethylene reactors with service temperature on the order of 1,000°C (1,830°F).[197] The coating is useful in preventing catalytic formation of soot and "metal dusting" leading to wall disintegration. In recent work, Ishikawa et al. evaluated multilayer approaches for preventing attack of type 316 stainless steel in contact with $MgCl_2$ at 700°C (1,290°F), an environment associated with titanium refining.[198]

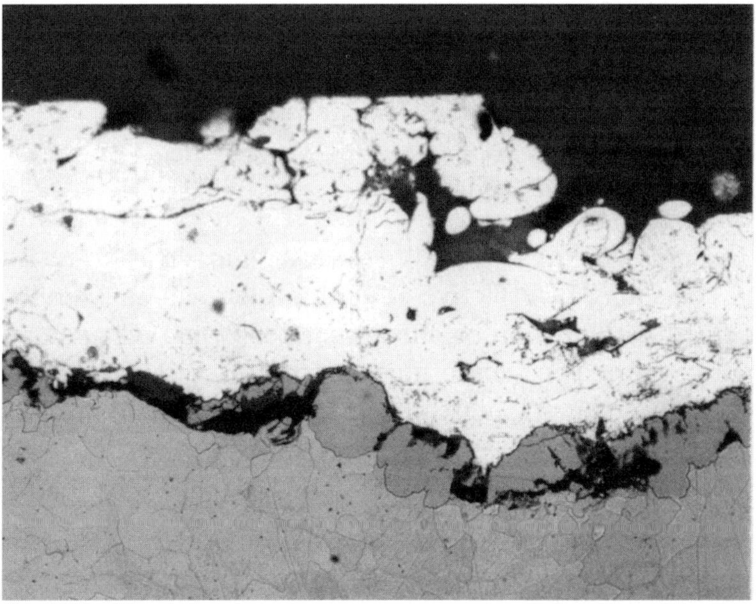

Figure 16.1b Thermal conversion of arc-sprayed aluminum coating after one hour at 600°C (1,110°F). Note the partial disbondment at the steel/coating interface and the formation of intermetallic compounds. 500X. Courtesy of Intermet Technology.

The primary objective was to use the multiple layers to circumvent the need for sealers or elaborate processing to avoid the problem of porosity. In the laboratory phase of work, a flame-sprayed, three-layer structure (Ni-20% Cr/Ni-15% Cr-bal. Fe/Al exterior) provided best performance, which the authors ascribed to the barrier effect of a diffusion layer between the aluminum and the middle layer. In actual service, plasma-sprayed Ni-20% Cr/Ni-45% Cr/TiO_2 showed excellent performance.

Oki et al. assessed the performance of Ni-Cr-Al coatings on austenitic stainless for high-temperature oxidation resistance.[199] Air plasma-sprayed prealloy powder coatings were compared to coatings derived from powder blends of Ni-20% Cr and aluminum. Superior performance was noted for the mixed powder coating, and this was equated with the influence on phase changes occurring during high-temperature exposure.

Ceramic Coatings

Ceramic coatings, generally used with a corrosion-resistant bond coat, may be effective when high-temperature service is combined with erosion or other wear problems. Chandler and Quigley reported on the usefulness of plasma-sprayed aluminum oxide in protecting against erosion around access doors in the superheater area of a coal-fired utility boiler.[188] Czechoslovakian workers have described several practical applications for Al_2O_3 coatings deposited using their water-stabilized plasma-spray system.[61] The coating protects the welded joints of covers at pipe inlets in recuperators exposed to attack by combustion products with service temperature up to 1,050°C (1,920°F). It is unclear if either of these specific applications involved use of a bond coat.

For further information on ceramic coatings, see Chapter 14 section on case histories involving combined wear and corrosion.

Rohm and Haas reported an unsuccessful application of plasma-sprayed chromium oxide on Illium† 98 for a sulfuric waste acid burner nozzle. Coating failure was equated with thermal shock problems.[200]

Environment-Associated Cracking at High Temperature

Japanese researchers reported that aluminum on type 304 stainless steel, if properly converted to aluminum-iron compounds by heat treatment, is effective in mitigating stress corrosion cracking problems.[197]

The practical usefulness of the Amoco Ultracoat technology has been demonstrated in the remedial coating of hydrocracker reactor vessel at an Amoco facility.[68] These 2.25% Cr-1% Mo steel vessels have an interior overlay of type 347 (UNS S34700) stainless steel to contend with exposure to hydrogen sulfide at high temperature and pressure. An improper heat treatment rendered the overlay prone to severe cracking, which was only aggravated by the heat associated with weld repair. Preliminary crack repair simulation tests demonstrated that Ultracoat type 316L stainless steel was protective, while a broad range of flame-, plasma-, and Jet-Kote† HVOF-sprayed coatings were not. This was followed by extensive repairs on two heat exchanger channel covers. Coatings showed successful performance when inspected after two years of service.

† Trade name.

CHAPTER 17

Applications — Protection of Steel from Atmospheric Corrosion

Thermal sprayed zinc and aluminum have long track records as protection strategies for structural steel, with documented case histories of bridge coatings applied in the 1930s that are still preventing corrosion. Economical spraying of large surface areas demands use of high deposition rate, high deposition efficiency processes such as wire combustion and especially arc-spraying. Because zinc (and aluminum in marine environments) is galvanically sacrificial with respect to steel, porosity is less critical than for other corrosion applications. However, a simple sealing step can greatly extend coating service life and overall economics.

In severe environments, more advanced coating compositions and/or combination with advanced paint topcoats may be appropriate. Zinc-aluminum prealloy spray wire offers some advantages in terms of corrosion response, as does co-spraying elemental aluminum and zinc. An aluminum-aluminum oxide wire feed is available for applications where abrasion is also an issue. Zinc-nickel coatings have been explored for protection in cases where stress corrosion cracking is a problem. Thermal spray has also been developed as the basis of protection strategies for steel reinforcement of concrete.

Thermal spraying of plastics has not been extensively documented. Recent literature indicates that chronic problems with adherence and performance can be overcome, so more widespread use can be expected.[104, 105]

Experience with Zinc and Aluminum Coatings

The most complete study of thermal spray coatings for corrosion protection of steel was initiated by the AWS in 1950. This extraordinary effort involved exposure of over 4,000 sealed and unsealed flame-sprayed coatings to industrial and marine environments (including near-shore, full and half-tide immersion in quiet sea water, and full immersion in flowing sea water). Documentation of 19-year exposure results indicated the following key conclusions:[112] 3- and 6-mil (76- and 152-µm) aluminum coatings, sealed and unsealed, gave complete protection in all environments. Curiously, thicker coatings did not perform as well because of the formation of pits and blisters. Damaged areas of aluminum-coated panels appear to be effectively protected by galvanic action.

Unsealed zinc-sprayed panels required 12 mil (0.3 mm) of coating for 19-year protection immersed in seawater and 9 mil (0.2 mm) for severe marine and industrial exposures. Proper sealing extended the life of thinner coatings to the 19-year evaluation point.

For atmospheric exposure, a very successful sealing strategy was an acid-zinc chromate wash primer followed by an aluminum flake loaded vinyl copolymer air-drying sealer. A clear vinyl copolymer was also used successfully for aluminum in seawater exposures. Chlorinated rubber was not satisfactory as a sealer for either zinc or aluminum.

Further testimony for the long-term protection afforded by aluminum in marine environments is a summary of 34-year performance of similar panels at the test site at Kure Beach, North Carolina, indicating that 3- and 6-mil (76- and 152-µm) coatings were still protective.[202] It is unclear if these were sealed panels.

The successful use of zinc-spraying to protect bridge structures has been documented,[203] as has the United States Navy's extensive utilization of aluminum arc-spraying.[4] In general, aluminum is preferred for marine exposures and slightly acidified industrial conditions. Zinc is recommended for rural and alkaline industrial exposures and for nonacid fresh water immersion below 140°F (60°C). Zinc is also appealing because it is more forgiving than aluminum with regard to less-than-ideal surface preparation.[204] Oxidation and embrittling phase reaction with the substrate restrict zinc use to relatively low service temperatures (i.e., 140°F [60°C]).[205] Aluminum, on the other hand, forms a tenacious protective oxide film at high temperature, making it suitable for applications such as exhaust stacks. Diffusion reactions yield desirable iron-aluminum phases similar to those formed in calorizing. This coating structure may, in fact, be intentionally generated by a post-coating heat-treatment. Successful use of sprayed aluminum up to 900°C (1,650°F) has been documented.[195]

Besides the vinyl copolymer sealers alluded to in the AWS study, successful sealers for zinc and aluminum have included various low-viscosity formulations based on phenolic, modified epoxy-phenolic, and polyeurathane resins.[38] For sealing aluminum in elevated-temperature applications (up to approximately 480°C [895°F]), an aluminum-pigmented silicone is appropriate.[195] United States Navy procedures for aluminum arc-spraying of off-shore platforms include five coats of sealer plus top-coat.[205] Payne dismisses the need for painting sealed thermal spray coatings for civil applications.[206] However, the corrosive environments of some plant applications may justify top coating. Elayaperumal identifies epoxy and alkyd paint systems for use in protecting structurals in the pulp bleaching section of a pulp and paper plant.[207]

There is documentation of successful applications for zinc and aluminum with specific relevance to the chemical process, petrochemical, and pulp and paper industries. Echoing the positive naval experience, offshore platform facilities (including both structurals and storage tanks) have been successfully protected with sprayed aluminum.[208, 209] Conoco's experience in this regard indicates that the aluminum coating is much more robust than paint systems in withstanding the rigors of construction, transport, and assembly. Elayaperumal alludes to aluminum-sprayed recovery boiler chimney stacks and recommends the use of sprayed aluminum or zinc with the appropriate paint top-coat for avoiding the deterioration of structurals in the pulp bleaching section of a pulp and paper plant.[207] Herman and Altorfer mention several zinc coating applications, including wellhead assemblies, offshore drilling rigs, flare stacks, refinery columns, oil and propane gas storage tanks, and railroad tank cars, both inside and out.[203]

When considering zinc coating applications, however, one needs to keep in mind a major caveat. Contact of zinc with austenitic stainless steel can lead to catastrophic liquid metal embrittlement in the case of fire or subsequent welding.[116] Zinc-coated materials should be isolated from stainless steel to avoid this possibility. Molten aluminum contact with austenitic stainless steel is also associated with corrosion problems, though it is unclear if actual liquid metal cracking can be triggered without high residual stress.[210] An aluminum coating on type 304 (UNS S30400) stainless steel, when properly converted to high-melting iron-aluminum compounds by heat treatment, has been reported to significantly upgrade resistance to stress corrosion cracking.[201] Another issue that may come into play when selecting coating composition is the tendency of aluminum coatings to spark.

Thermal spraying of structural steel offers some distinct advantages over the competitive coating strategies. There is no size limitation as in the case of hot dip galvanizing. The possibilities of thermal distortion, cracking caused by thermal stresses, hydrogen pick-up during pretreatment, and flux inclusions are avoided. Enhanced protection may be introduced to local areas prone to severe corrosion, such as splice plates and embedments in concrete. This strategy has recently been promoted for protection of structurals in pulp and paper mills.[211] Compared to advanced paint systems, the economics of thermal spray may be advantageous when maintenance costs and life of coating are considered.[205]

It is noteworthy that submerged steel structures can be successfully coated by underwater spraying. Olinger et al. reported on underwater sand blasting followed by HVOF spraying of zinc, aluminum, and alloy 625 (UNS N06625).[212] Good bonding was attained for all these coatings.

Recent Innovation in Feedstock Materials

Another potential advantage of thermal spraying is a great deal of flexibility with regard to coating composition. For corrosion protection of steel, work on innovative alloy development has been largely confined to the zinc-aluminum system. A good deal of technical work has been performed since the 1960s,[213] with the focus of commercialization being zinc-15 wt% aluminum wire.[214] In general, laboratory evaluations have indicated superior corrosion response compared to elemental zinc or aluminum, and vendor product literature emphasizes a dramatic increase in anticipated coating life in high-SO_2 atmospheres. However, severe blistering problems have been noted in marine splash and spray exposure, as well as in saltwater immersion.[89] Evaluations of multilayer zinc-on-aluminum and aluminum-on-zinc coatings have also been performed, with somewhat ambiguous results. Another zinc-aluminum coating approach has been dual wire arc-spraying, which involves feeding zinc and aluminum wires simultaneously. Shaw et al. indicated somewhat encouraging results in marine environments for coatings with 15 wt% aluminum,[89] while Zhongli et al. focused on 28% Al, which corresponds to a 1 to 1 volume ratio.[215] In this research, outstanding performance has been exhibited in saltwater immersion and moist SO_2 atmosphere.

Another recent commercial innovation in feedstock has been the Duralcan† 90/10 metal matrix composite. This aluminum-based wire suitable for flame- and arc-spray contains a fine dispersion of 10 vol% aluminum oxide. Compared to pure aluminum, the composite coating exhibits a 10-fold enhancement in wear resistance (thrust-washer test) and a 35 percent improvement in abrasion resistance. Corrosion response is comparable to that of aluminum. The nonslip/nonskid character of thicker arc-spray coatings suggests usefulness in industrial applications such as pedestrian walkways, ladders, scaffolding, and vehicle ramps.[216]

Coatings such as zinc (and aluminum, in some environments) protect exposed steel surfaces by sacrificial galvanic action. This is generally an advantage, but it can cause problems when higher-strength substrates are used. This is because the galvanic potential difference drives the formation of hydrogen at the cathodic surface, which in turn can trigger hydrogen-associated embrittlement phenomena such as stress corrosion cracking.

Recent research has explored zinc-nickel powder combustion spraying as a protection strategy for high-strength steel.[91] A coating with 18% Ni is still sacrificial with respect to steel, but the galvanic potential difference is not sufficient to drive formation of hydrogen at exposed steel surfaces. Zinc-nickel evolves a corrosion product that is much more dense and compact than that associated with zinc. Coating pores will not be plugged with corrosion product, making sealing more important than for conventional zinc. Coating adhesion and brittleness on impact are

† Trade name.

problems, and research to enhance coating mechanical response is being pursued.

Dow Chemical workers documented a significant advance in the area of thermal sprayed plastics.[105] Ethylene-acrylic acid (EAA) copolymers can be sprayed to yield highly adherent coatings, *if* proper attention is paid to control heat input so that excessive cross-linking does not occur. Most of the documented case histories for these coatings have been in site-applied civil applications (light poles, bridge rails, and supports), but use in contact with hydrogen peroxide and amine has been alluded to. Dow indicates that the copolymer is resistant to a broad range of chemicals, solvents, and reagents. Some exceptions include chlorinated solvents, fuming or strongly oxidizing acids, aromatic alcohols, and heterocyclic aldehydes. Service temperature is limited to 165°F (79°C). National Coatings has extensive experience with field applied plastic thermal spray and indicates that the DuPont ethylene-methacrylic acid (EMAA) dry-blend powder also gives outstanding performance.[218] Demanding commercial applications include salt hoppers, fertilizer handling equipment, and sewage treatment facilities. Literature from UTP indicates that ethylene/vinyl alcohol copolymer is appropriate for long-term corrosion protection of steel. Its availability in a wide range of colors is useful for pipe identification in chemical plants. UTP recommends its SuperPlast[†] nylon feedstock and EAA for more severe environmental exposure.

Protection of Steel Reinforcement in Concrete

Thermal spraying is also playing a role in evolving strategies for the prevention of corrosion of steel reinforcement in concrete. This is a key issue in infrastructure rehabilitation and has relevance to CPI applications, as well. While steel is normally passive in the alkaline concrete environment, chemistry shifts such as chloride intrusion and carbonation reactions can trigger active corrosion. The high specific volume of rust causes cracking and spalling of concrete and further access of corrosion-promoting species.

The recent focus of thermal spraying for this application has been zinc deposition on concrete surfaces to serve as the anodes in cathodic protection systems.[219] In general, these have been active, impressed-current systems. However, highly conductive concrete may permit the "passive" approach, which relies on the natural galvanic potential difference between steel and zinc. After removal of unsound concrete, exposed reinforcing bar may be blast-cleaned, sprayed, and left exposed to the environment.

For new construction, another alternative is coating the rebar itself. Commercial options (not thermal sprayed) include fusion-bonded epoxy and hot dip galvanizing. The former is currently prevalent, although a rash of field failures has triggered renewed interest in zinc-based approaches. A problem with galvanizing is that zinc evolves hydrogen bubbles when it first contacts wet cement, causing a degradation of the coating/cement interface. An environmentally undesirable chromate treatment is required to control this gassing reaction. Thermal spraying offers the flexibility of feedstock selection to circumvent this problem. Specifically, Krepski has described a zinc-silicate composite spray for the rebar application.[92] The zinc-nickel alloy mentioned above also exhibits no hydrogen bubbling when placed in electrolyte-simulating wet cement.[91]

CHAPTER 18

Applications - Heat Transfer

Plasma-sprayed ceramic coatings provide thermal barrier protection of metallic components in gas turbines and diesel engines, lowering operating temperature and improving operating efficiency.[38] Preferred coatings are "stabilized" zirconia, with yttria (8%) or ceria-yttria to provide long-term chemical and thermal stability.[99] An oxidation-resistant bond coating such as M-CrAlY is utilized.

Avoiding stress-induced coating failure caused by thermal shock and cycling is a key to good performance. To better match thermal expansion, graded coatings with intermediate oxide plus metal layers may be employed. A key problem with these coatings is vanadate corrosion caused by vanadium-containing fuels. Yttria reacts with molten vanadate and triggers coating spalling.[220] Alumina may also be used in heat shield applications.[99] Coating porosity may, in some cases, be an advantage in restricting heat transfer.[83]

Magnesium zirconate is a nonwetting thermal barrier coating used in the containment and handling of molten materials.[38] Research has indicated that the coating is more tolerant to thermal shock if a fused and quenched powder is sprayed rather than a fused, slow-cooled, crushed powder.[71] This may relate to the fully cubic crystal structure in the former case, while the slow-cooled powder contains some monoclinic phase. Vestiges of this difference are apparently retained in the product coating.

Berndt et al. have reported on the usefulness of sprayed blends of zirconia or alumina with Ni-Al for short-term fire protection of aluminum structures.[221] Best results were obtained with the 20% alumina and 50% zirconia cermet compositions.

The Japanese steel industry has done extensive work on repair of refractory furnace linings by a "flame gunning" technique.[222] This is essentially oxy-propane flame-spraying with a relatively coarse particle size. Depending on refractory liner and furnace environment, the feedstock may be silica-calcia, alumina-silica, alumina-magnesia, magnesia-calcia-silica, or magnesia-alumina-calcia.

In the pulp and paper industry, an important application for thermal spraying is the resurfacing of Yankee dryer rolls. While this subject fits under the previous categories of "restoration and repair" and "wear," efficient heat transfer is also critical. Hayes and Thun of CWS documented the historical evolution of coatings for Yankee dryers.[183] First experience was with combustion-sprayed type 402 (UNS S40200) stainless steel applied to edges only to contend with doctor blade wear. This coating required a bond coat. Because it was usually applied by mill personnel, quality was not consistent. The next stage in this evolution was arc-spraying, which eliminated the need for a bond coat. This and the significantly higher deposition rate meant less down time. Spraying by the applicators' coating service personnel rather than in-house staff yielded more consistent coatings, and full face coatings became the accepted standard. The main problem with this coating was poor heat transfer, resulting in a 10 to 15 percent loss in production.

CWS addressed this market need in a four-year development program involving plasma-spraying with a novel molybdenum-based feedstock containing small concentrations of chromium and nickel. This composition was selected based on the coating's excellent thermal conductivity. Heat transfer is still cut four percent compared to the uncoated surface, but overall production is improved because of the excellent wear performance of the coating. Wear rate is reduced five-fold in some cases. This coating has given three years of service. Occasional problems are associated with unusual stress patterns, edge cracking, and corrosion associated with extremes in pH.

An exquisite illustration of the potential of thermal spray coating to improve process efficiency in the pulp and paper industry has recently been provided by Lenling, Smith, and Orloff.[223] In impulse drying, water is removed from the paper by the propagation of steam pulse, which physically expels the water as the paper passes between the nips of a hot press roll. This is far more energy efficient than drying by simple vaporization, and it is estimated that $800 million could be saved by widespread implementation of this technique in the United States alone. A major problem with impulse drying is paper delamination, which is believed to be caused by conversion of superheated water within the paper to steam at the exit of the high-pressure region of the roll "pinch." By restricting heat transfer from rolls to paper, the delamination problem can be minimized. This was accomplished in the reported research by using a plasma-sprayed arc-fused/spray-dried, partially stabilized zirconia coating (Zircoa Zirspra† 9204), with spray parameters adjusted to give very high (17 percent) porosity. A thin (2-mil [0.05-mm]) top layer of more dense zirconia was introduced to prevent excessive escape of steam through the porous coating, which would otherwise reduce the intensity of the steam pulse. A Ni-Al-Mo bond coat was employed. Results of both laboratory simulation and pilot-scale roll press experiments demonstrated successful impulse drying without paper delamination. A full-scale commercial test is anticipated before the end of 1995.

† Trade name.

CHAPTER 19

Special Applications — Repair of Glass-Lined Vessels and Spraying of Silicate Glass

APS-Materials markets a plasma-sprayed coating approach for on-site repair of glass-lined vessels, called Ta-Coat.[†148] The technique is generally only recommended when the standard mechanical plug approach is not appropriate—at nozzles and complex curved surfaces, for example. A relatively thin (less than 8 mil [0.2 mm]) tantalum basecoat is followed by a topcoat of chromia. An organic sealer may also be used, depending on the process environment. The technique may not be successful if the substrate has been hydrogenated under strongly oxidizing conditions. Diamond Shamrock has reported on implementation of this technology to repair damage at the manhole, along baffles, and at the bottom of a 2,000-gal. (50,000-L) vessel used in production of herbicide/fungicide.[152] Relining this vessel would have cost $45,000 and taken several months. The repair procedure cost $13,000 and took two days. After three years of service, the repair showed no evidence of problems.

Berndt has reported on work by Arata et al. on plasma-spray repair of glass-lined vessels with alumina and rutile (i.e., titania).[224] The key to successful coating is keeping the low thermal conductivity silicate glass substrate temperature in the range of 500 to 600°C (930 to 1,110°F) during spraying by control of powder feed rate and spray distance. If the glass substrate temperature gets too high, thermal shock problems can lead to cracking. Coating thickness up to 0.3 mm (12 mil) without the use of any bond coat is reported.

[†] Trade name.

Noji et al. performed pioneering work on flame-spraying of silicate glasses onto steel.[225] Adequate substrate preheat and a long standoff for adequate flame residence time of the feedstock are critical to success. More recently, Surface Alloying and Wearfacing Services of Houston has been offering a Crystallite† sprayed glass coating, but no specific case history information is available.[179]

Rough Wear-Resistant Coatings for Gripping

The paper industry requires coatings for applications such as winding cylinders that maintain a rough surface profile in service for gripping of product. Polak and Witts reported on experience in this area.[226] Flame-sprayed molybdenum did not retain its rough surface after eight months of service. Best results were obtained for a nickel-based alloy reinforced with tungsten carbide over a Ni-Al bond coat. The selective wear of the mixed hard and soft phases in the coating appears to be a key to maintaining the rough profile.

Recently, Scholl and Clayton reported on experimental work aimed at improving the efficiency of the pulping process by changing the texture of refiner plates via high-energy plasma-spraying.[170] By manipulating powder particle size and spray distance, surface roughness was varied for several different metal, oxide, and cermet coatings. There was no correlation between surface roughness and abrasive or erosive wear.

The Duralcan† nonskid aluminum-aluminum oxide coating deposited by arc spraying could be an attractive low-cost alternative for this type of application.[216]

Clearance Control Coatings

In gas turbines, it is desirable to maximize flow over blade surfaces and minimize loss between the blade tip and casing liner.

The casing liner is coated with an "abradable" composition, while the blade tip receives an abrasive coating. The rotating blade generates a small channel in the liner, providing minimal clearance and maximum sealing.[227]

Besides accommodating the abrasive action of the blade, the abradable coating needs to have high gas and particulate erosion resistance, good thermal stability, and self-lubrication. For the compressor section, typical abradable coatings include nickel or aluminum plus graphite, aluminum-silicon plus polyester, nickel-chromium plus polyester, polyeurathane, or bentonite. For the hotter turbine section, zirconia-yttria or zirconia-magnesia with an M-CrAlY bond coat applied by air plasma-spray is typical.

The abrasive coating for the blade needs to wear the channel in the abradable material at an acceptable rate without much heat generation and needs to exhibit good wear resistance. For the compressor section, alumina or nickel-chromium loaded with chromium carbide is utilized. In the turbine section, M-CrAlY loaded with additions having an abrasive character may be used.

Antifouling Coatings

Copper and Cu-Ni arc-wire sprayed coatings have been used to prevent biofouling for offshore platforms.[228] An innovative approach to biofouling is offered by Copperlok Inc. of Hatboro, Pennsylvania.[229] An epoxy bondcoat is first deposited by conventional roller, brush, or spray technique. Before this coating cures, hollow glass spheres are bonded to the surface by a special spraying system. After the epoxy cures, the surface is lightly abraded to break the spheres. The surface is then wire flame-sprayed with Cu-Ni, and the interlocking with the broken spheres establishes a strong bond. This technique has been applied to fiberglass, steel, ferrocement, and wooden ship hulls; and probably has more general applications, as well.

CHAPTER 20

Practical Aspects of Coating Utilization

Defining the Mode of Deterioration

Successful implementation of a thermal spray coating will be more likely if the demands on the coating are fully understood. Different wear modes require different materials strategies—the old saw of "harder is better" only holds true in limited circumstances. The complex environments encountered in the CPI can give rise to corrosion mechanisms dominated by the roles of low-level impurities. Thorough analysis of uncoated components and previously utilized coatings that have failed in service is invaluable in providing a baseline on which to build. Reference materials such as the MTI *Atlas of Corrosion and Related Failures*,[230] NACE publications on corrosion in the CPI,[116, 231] and the ASM International[(9)] *Metals Handbook* volumes "Corrosion"[232] and "Tribology"[233] are useful references for both defining deterioration mode and identifying possible candidate materials.

Selecting Candidate Coatings

Once the performance environment is defined, the published literature and past performance of bulk materials should help establish candidate coatings. Experienced applicators also have

[(9)] ASM International, Materials Park, OH.

a wealth of unpublished case history information that might be relevant to a given application. Although the catalogues of feedstock suppliers are continually expanding, available compositions and particle size ranges are limited. Several companies offer toll atomization services, should exploration of new alloys be warranted. Special composite powders can also be produced in small batch runs. Microstructural effects may result in coatings that behave quite differently from their wrought or cast counterparts, and different forms of the same feedstock composition may yield different results.

If a bond coat is required, care must be taken in selecting the appropriate option for a given environment. Likewise, choice of a sealer should not be an afterthought—key issues are chemical and thermal stability and the quality of the bond between sealer and coating surface. The more sophisticated suppliers should have specific recommendations. If practical, the option of vacuum impregnation with sealer should be considered.

Ceramic coatings are often specified for wear resistance, high-temperature stability, and thermal barrier performance. Thermal expansion data can be useful for initial screening of candidate coating/substrate combinations tolerant of thermal shock and cycling. The user needs to recognize that generally poor fracture toughness means these coatings will not withstand rough handling or impact during service. The cost of cobalt-based alloys such as Stellite† 6 is of concern because of unstable world supply for this strategic metal. Iron- and nickel-based alternatives should be considered. If the coating is to receive high bearing loads, compositions amenable to spray-and-fuse or plasma transferred arc methods may be appropriate.

Selecting the Spraying Process

Once the feedstock is decided upon, selection of the coating system may, to some extent, be a trade-off of coating quality and

† Trade name.

economics. Much of the initial interest in HVOF technology has been based on its potential as a lower-cost alternative to the premium D-Gun tungsten carbide cermets. Chamber plasma-spraying can yield the reduction of porosity needed for some corrosion-related applications. However, size limitation and the cost of automated spraying of complex shapes may make it impractical for many industrial uses. Arc-spraying with inert-gas atomization and HVOF with gas shrouding are more realistic options for industrial applications, if reduction of porosity is critical.

For economical coverage of large surface areas, wire arc-spraying is the preferred delivery system. The limited selection of ductile wire feedstocks is being supplemented by availability of cored wire materials.

Air plasma-spraying is the workhorse system for deposition of oxide ceramics. The Norton Rokide[†] rod-fed system is a low-capital cost alternative if an in-house capability is to be established.

Laboratory Testing vs Plant Trials

For some evaluations of new coating approaches, it may be appropriate to go directly to testing in actual service. For instance, upgrading the wear performance of an uncoated component can often be explored without serious negative implications if the coating fails. This is not the case for many corrosion-related applications, especially if a coating on a low-cost substrate is being considered as an alternative to a more expensive bulk material. A laboratory screening evaluation is a prudent first step, but only if the test environment carefully simulates actual service. Corrosion testing needs to address pH variation, low-level process solution impurities, flow conditions, and thermal cycling. Stress conditions should be carefully considered, especially if mechanisms such as stress corrosion cracking may be involved. Laboratory standard wear tests may be useful in comparing and optimizing coatings, but caution should be exercised in extrapolating the results to actual service. If an application involves only local coating, edge effects and galvanic coupling need to be considered. Experimental design should be statistically valid to ensure that comparisons of performance are meaningful. Expected variability in coating character and effects of localized damage should be considered; the "robustness" of process and coating are often critical to the success of a coating. The testing of coupons or patches in the actual service environment may be appropriate initial or intermediate steps.

In-house Capability vs Purchased Service

Some of the early field failures of thermal spray coatings may be traced to less-than-adequate appreciation of process control and quality control issues by in-plant personnel. The experience base of a dedicated applicator helps to ensure more reliable coatings and is also a valuable resource in developing a coating strategy for a new application. However, the input of plant personnel who are intimate with the service requirements is also critical. Some companies providing coating services tend to keep the details of their proprietary technologies to themselves. The chemical process industries present such a range of service environments that a high level of technical dialogue between the coatings expert and the plant engineer is needed.

If the situation justifies the establishment of an in-house coating capability, effective training of applicators is important. With the growing interest in thermal spray over the past few years, training resources are becoming more available. Particularly noteworthy are the courses offered by ASM International and the training "institutes" established by Hobart-TAFA, Eutectic-

Castolin, and Metco. Some applications for thermal spray require more sophistication than others. For instance, restoration of worn mechanical components using wear-resistant, self-bonding powder is a good candidate for an in-house spray capability. The spray process is relatively robust and forgiving, and the simple powder-fed combustion torch used represents a relatively low-capital cost. A good example is the Wear Management Service Tribolite† "One-Step" iron-based powder blend.[7]

Shop vs Field Applied

For specific market niches like large rolls, boilers, and process vessels, coating shops have established field-application services that will considerably reduce down time. Spray systems include combustion, plasma, and HVOF (including the Amoco Ultracoat† technology). Applications may involve automated control of spray torch position. Of course, protection of steel structurals by thermal spray is largely by on-site application. Shop-applied coating of new heavy structurals would appear to be a viable alternative to hot-dip galvanizing.

Conclusion

Thermal spray coating may not be the answer to every problem involving degradation of a component surface. It is just one weapon in an arsenal that includes many other coating strategies: cladding, inhibitors, cathodic protection, and alternate bulk materials. Thermal spraying *does offer* a great range of coating options, and provides an appealing opportunity for innovation. Some of the possibilities are addressed in Chapter 21.

CHAPTER 21

Future Trends in Thermal Spray Coating for the CPI

Recent advances in thermal spray coating technology justify continued evaluation of its usefulness in solving materials problems confronting the CPI. Such efforts to upgrade materials performance are more likely to succeed if the focus is on the character of the coating itself. Simple designation of coating composition and method of application does not define performance. An array of process variables must be controlled to ensure coating quality. Even if optimal deposition parameters are attained, the coating cannot be considered equivalent to a bulk wrought component or clad product. The engineer responsible for materials selection need not be a trained materials scientist, but should at least be aware of these issues.

Several emerging themes that have been discussed in the text have specific relevance to the CPI and merit a closer look by engineers responsible for materials innovation. These include:

- Vacuum or inert-atmosphere chamber spraying (including use of the process vessel as the chamber);
- Arc-spray with inert-gas atomization and shrouding;
- HVOF with inert-gas shrouding;
- Spraying of tantalum, zirconium, titanium, and molybdenum;
- Innovative corrosion-resistant bond coats for ceramics;
- Ceramics with enhanced fracture toughness;
- Silicate and oxide glasses;

- Intermetallic compounds such as aluminides for enhanced high-temperature environmental response;
- Amorphous metals and alloys designed based on rapidly solidified microstructure;
- Sprayed polymers;
- Reactive sealers;
- Multilayers that form dense layers by heat treatment;
- Feedstock blends containing bond-enhancing powders; and
- Coatings to prevent environmentally assisted cracking.

Coating porosity is a primary consideration for corrosion protection, and a major thrust of recent research and development has been minimization of porosity. Chamber spraying has yielded desirable coatings in this regard, but high cost and size limitations will continue to limit utilization. The concept of spraying vessel interiors using the vessel itself as a controlled atmosphere chamber may become viable as systems for automated spraying become more refined. Inert-gas atomization and shrouding appears to be a means of upgrading the quality of arc-sprayed coatings, and the range of feeds available for this low-cost, high deposition rate process may be extended through the use of the cored wire approach.

With respect to cost-effective upgrading of coating quality, the most exciting recent advances have probably been associated with refinements of the HVOF approach, including inert-gas shrouding. Such technologies have yielded very encouraging results in preliminary industrial applications. Particularly noteworthy is a relative insensitivity to deposition angle, which has great relevance to the protection of complex profiles such as boiler tube walls. For corrosion applications where no through-porosity is permissible, it will be important to develop meaningful quality assurance procedures to verify acceptable coating structure.

The reactive metals tantalum, zirconium, titanium, and molybdenum offer attractive corrosion response in many hostile environments. However, sensitivity to atmospheric interactions has been a problem in commercial utilization of thermal spray coatings for these metals. Controlled atmosphere spraying and other refinements in processing may provide the means to avoid this difficulty.

Ceramic coatings merit greater consideration for applications involving corrosion plus wear. In identifying the coating system for a specific chemical environment, as much care needs to be taken in choosing the bond coat as in selecting the top layer. Besides coating porosity, the generally poor fracture toughness and fragility of ceramic coatings makes it advisable to rely on the bond coat as the primary corrosion barrier. The workhorse nickel-aluminum bond coat is often specified inappropriately. Considering the early use of molybdenum as a bond coat and its attractive corrosion response, it is surprising that more use has not been made of this material. Another approach is to deposit the bond coat by a technique other than thermal spray to ensure a continuous corrosion barrier. For instance, electrochemical techniques exist for the deposition of continuous tantalum films on ferrous substrates. A multilayer coating of plasma-sprayed oxide over tantalum would overcome the poor wear response of tantalum while taking advantage of its corrosion resistance.

One route to improved ceramic coatings would be to take advantage of the materials science innovations responsible for the new wave of remarkably tough structural ceramics. Approaches such as fiber loading, controlled microcracking,[234] and reaction toughening could yield significant improvements in coating response. The relatively simple concept of blending oxide feeds with metal powders has only been cursorily explored for corrosion plus wear applications. Another family of materials that has not received much attention is silicate and other oxide-based glasses. Glasses have good stability in severe environments, as attested to by the extensive use of glass-lined vessels. Modern glass-ceramics display outstanding fracture toughness and resistance to thermal shock. The rapid quench associated with thermal spraying can yield novel amorphous or partially crystallized structures in systems such as alumina-yttria and Cordierite.

Another emerging family of materials is that of intermetallic compounds such as nickel and iron aluminides. Brittleness problems for these materials are being overcome, and the feedstocks used for manufacture of components through powder metallurgy techniques could also be used in thermal spraying. Aluminides are especially attractive for applications where high-temperature environmental response is the key issue.

In general, the importance of rapid solidification needs to be more fully appreciated in the design of thermal spray coatings. Coating alloy compositions are still largely based on cast- and wrought-alloy developments. Compositions should be based on the microstructure that evolves under the rapid quenching conditions of spraying. Amorphous metal coatings such as the Metco[†] 700 family provide properties different from their crystalline counterparts. However, the limited thermal stability of these microstructures needs to be recognized.

Considering the extensive utilization of extruded and injection-molded polymer components in harsh chemical environments, surprisingly little effort has been devoted to the possibility of developing polymer coatings through thermal spray. With proper substrate preparation and control of temperature, the new EAA and EMAA resins reportedly avoid the adhesion problems previously associated with sprayed plastics. Coatings may be applied on site and do not have the problem of volatile organic components.

Development of novel sealers provides an under-utilized route to coating improvement for CPI applications. At present, sealing is for the most part either an "afterthought" paint-on or a high-tech, high-cost process like laser glazing or hot isostatic pressing. Self-curing, water-based silicates and other ceramic paints should be more thoroughly explored. Sealers that interact with the coating to yield tightly bonded, dense reaction products in pores would be highly desirable. Another approach to sealing is the spraying of multiple layers that react to form dense barrier layers during solid-state heat treatment.[198] The option of blending bond-enhancing metallic powders into other feedstocks should continue to be explored. This could be a means of improving bonding of splats to each other as well as to the substrate.

Stress corrosion cracking and hydrogen-induced cracking are two chronic problems in the CPI that could be alleviated by thermal spray coatings. Among the approaches deserving further research are coatings that provide galvanic protection without evolution of hydrogen as part of the cathodic reaction and dense intermetallic layers based on the reaction of aluminum with the ferrous substrate during heat treatment.

Another key is a more complete understanding of the process environment and its implications with regard to materials selection. The CPI has numerous examples of low-level chemical species that dramatically impact materials response. For instance, oxidizing impurities in chloride process streams can evoke a highly detrimental pitting response to stainless steel, and molybdenum additions to the alloy are known to mitigate the problem. The engineer responsible for materials selection needs to access available technical information of this kind as part of the decision-making process.

Field tests of thermal spray coatings have sometimes yielded disappointing results. Hopefully, this review puts such experience into perspective. Coating structure determines coating performance and is, in turn, dependent upon numerous process variables and the exact nature of the feed material. As these relationships are better understood, the success rate for thermal spray coating applications will improve. Advances in process control, including closed-loop expert systems, in-process quality control, and diagnostics, will help to attain more consistent high-quality coatings. With ongoing refinement of spray systems, premium coatings will become more cost effective. Thermal spray coating is an option that merits careful consideration when tackling the challenging materials performance problems of the CPI.

APPENDIX A

1993 Thermal Spray Buyers' Guide

This listing of providers of thermal spray products and services was originally published in *Advanced Materials and Processes* 143, 5 (May 1993). It is reproduced with the kind permission of ASM International.

1993 THERMAL SPRAY BUYERS' GUIDE COMPANY DIRECTORY

A

A-Flame Corp.
8175 Kroger Farm Rd.
Cincinnati, OH 45243
Tel: 513/831-4284
Fax: 513/831-5237

A & A Co. Inc.
2700 South Clinton Ave.
South Plainfield, NJ 07080
Tel: 908/561-2378
Fax: 908/561-4201

A B Seals Inc.
321 S. Hamilton St.
Sullivan, IL 61951
Tel: 800/582-7992
Fax: 217/728-7580

Adiabatics Inc.
3385 Commerce Dr.
Columbus, IN 47201
Tel: 812/372-5052
Fax: 812/372-4470

Advanced Materials Technologies Inc.
2015 W. Alameda Dr.
Tempe, AZ 85282
Tel: 602/438-8760
Fax: 602/431-9051

Aerobraze Corp.—Coatings
940 Redna Terrace
Cincinnati, OH 45215
Tel: 513/772-1461
Fax: 513/771-1248

Air Products and Chemicals Inc.
7201 Hamilton Blvd.
Allentown, PA. 18195-1501
Tel: 215/481-5736
Fax: 518/686/4840

Alamo Supply Co.
1331 Upland Dr.
Houston, TX 77043
Tel: 713/932-8674
Fax: 713/932-7616

Allied Signal/Flourglas
14 McCaffrey St.
Hoosick Falls, NY 12090
Tel: 518/686-7301
Fax: 518/686-4840

Alloy Carbide Co.
P.O. Box 5368
Houston, TX 77262
Tel: 713/923-2700
Fax: 713/923-2700

American Metallizing & Sandblasting Inc.
90 Elinor Ave.
Akron, OH 44305
Tel: 216/794-0022
Fax: 216/794-1679

Ametek Inc.
Rt. 519
P.O. Box 427
Eighty Four, PA 15330
Tel: 412/225-8400
Fax: 412/225-6622

Amorphous Metals Technologies Inc.
27722 El Lazo Rd.
Laguna Niguel, CA 92656
Tel: 714/643-1700
Fax: 714-643-1125

Anval
301 Rt. 17N, Ste. 800
Rutherford, NJ 07070
Tel: 201/939-1065
800/99-Anval
Fax: 201/939-1608

A. O. Smith Corp.—Protective Coatings Div.
8160 Holton Rd.
Florence, KY 41042
Tel: 606/727-3500
Fax: 606-727-6776

APS-Materials Inc.
153 Walbrook Ave.
Dayton, OH 45405
Tel: 513/278-6547
Fax: 513/278-4352

Applied Powder Technologies Inc.
540 5th Ave. S.W., Ste. 1415
Calgary, Alberta T2P 0M2
Canada

Applied Polymer Systems Inc.
4302 E. 10th Ave. #404
Tampa, FL 33605
Tel: 813/247-3065
Fax: 813/247-5923

ASB Industries Inc.
1031 Lambert St.
Barberton, OH 44203
Tel: 216/753-8458
Fax: 216/753-7550

B

Bay State Abrasives —Div. of Abrasive Industries Inc.
12 Union St.
Westboro, MA 01581
Tel: 800/366-4431
Fax: 800/366-4311

B.C. Metals Inc.
4207 Forsythia Dr.
Cincinnati, OH 45245
Tel: 513/753-0840
Fax: 513/753-6940

Belmont Metals Inc.
330 Belmont Ave.
Brooklyn, NY 11207
Tel: 718/342-4900
Fax: 718/342-0175

Bender Machine Inc.
2150 E. 37th St.
Vernon, CA 90058
Tel: 213/232-2371
Fax: 213/232-1790

Blakely & Associates
503 Beacon St.
Alton, IL 62002
Tel: 618-462-0856
Fax: 618-474-2109

Blast-It-All—Div. of Larry Hess & Assoc. Inc.
185 Piper Ln.
P.O. Box 1615
Salisbury, NC 28145
Tel: 704/637-3300
Fax: 704/638-9311

Brodeur Machine Co. Inc.
62 Wood St.
New Bedford, MA 02745
Tel: 508/995-2662
Fax: 508/995-9824

F. J. Brodmann & Co.
2072 Sussex St.
Harvey, LA 70058
Tel: 504/340-9579
Fax: 800/825-5352

Buehler Ltd.
41 Waukegan Rd.
Lake Bluff, IL 60044
Tel: 708/295-6500
800/Buehler
Fax: 708/295-7979

C

C & P Surface Technology
4230 Industrial Pl.
Island Park, NY 11558
Tel: 516/889-4784
Fax: 516/889-0773

Cerac Inc.
P.O. Box 1178
Milwaukee, WI 53201
Tel: 414/289-9800
Fax: 414/289-9805

Cincinnati Thermal Spray Inc.
5901 Creek Rd.
Cincinnati, OH 45242
Tel: 513/793-0670
Fax: 513/793-4254

Cincinnati Thermal Spray—South
11766 N.E. Hwy. 210
Rocky Point, NC 28457
Tel: 919/675-2909
Fax: 919/675-0806

Control Vision Inc.
P.O. Box 51505
Idaho Falls, ID 83405
Tel: 208/523-5506
Fax: 208/523-5520

Climax Specialty Metals
21801 Tungsten Rd.
Cleveland, OH 44117
Tel: 216/692-3990
Fax: 216/692-0029

D

DCE Inc.
11301 Electron Dr.
Jeffersontown, KY 40299
Tel: 502/267-0707
Fax: 502/267-3074

Douglas Call Co. Inc.
5741 Bay Side Rd.
Virginia Beach, VA 23455
Tel: 804/450-3336
Fax: 804/363-9131

Drexel University Center for the Plasma Proc. of Metals
32nd & Chestnut St.
Philadelphia, PA 19104
Tel: 215/895-1844
Fax: 215/895-6760

Dynamet Powder Products
600 Mayer St.
Bridgeville, PA 15017
Tel: 412/257-5012
Fax: 412/257-5154

E

ECS Metals Ltd.
P.O. Box 42
Syosset, NY 11791
Tel: 516/922-1111
Fax: 516/624-7743

Electro Plasma Inc.
17426 Daimler St.
Irvine, CA 92714
Tel: 714/863-1834
Fax: 714/863-6906

Ellison Surface Technologies Inc.
1780 Anderson Blvd.
Hebron, KY 41048
Tel: 606/586-9300
Fax: 606/586-8585

Empire Abrasive Equipment Co.
2101 West Cabot Blvd.
Langhorne, PA 19047-1893
Tel: 215/752-8800
Fax: 215/752-9373

Engineered Coatings Inc.
65 Inwood Rd.
Rocky Hill, CT 06067
Tel: 203/257-1119
Fax: 203/257-1319

Engineered Powder Technology
120 Research Dr., Ste. E
Milford, CT 06460
Tel: 203/878-5790
Fax: 203/877-8014

Eutectic Corp.
40-40 172 St.
Flushing, NY 11358-2799
Tel: 718/358-4000
Fax: 718/445-1242

Exline Inc.
3256 E. Country Club Rd.
P.O. Box 1487
Alina, KS 67402-1487
Tel: 913/825-4683
Fax: 913/826-4425

F

Fisher Products Inc.
1320 West 22nd Pl.
Tulsa, OK 74107
Tel: 918/582-2204
Fax: 918/582-2026

Flame Spray Coating Co. Inc.
33847 Doreka
Fraser, MI 48026
Tel: 313/293-8600
Fax: 313/293-8607

Flame-Spray Industries Inc.
S.U.N.Y. - Stony Brook Heavy Engr.
Bldg. 211
Stony Brook, NY 11050
Tel: 516/632-8710
Fax: 516/632-8708

Flame Spray Northwest
1605-1 S. 93rd St.
Seattle, WA 98108
Tel: 206/762-4144
Fax: 206/762-8019

Flamespray Engineering & Research
1835 W. 1500 South
Salt Lake City, UT 84104
Tel: 801/423-4182
Fax: 801/977-9261

F.W. Gartner Thermal Spraying Co.
3824 Lamar
Houston, TX 77023
Tel: 713/225-0010
Fax: 713/229-9841

Furon—CHR Div.
3660 Edison Pl.
Rolling Meadows, IL 60008
Tel: 708/392-8090
Fax: 708/392-0403

G

General Magnaplate Corp.
1331 US Rt. #1
Linden, NJ 07036
Tel: 908/862-6200
Fax: 908/862-6110

General Plasma Inc.
5 Draper St.
Woburn, MA 01801
Tel: 617/933-9386
Fax: 617/933-0841

General Plasma Inc.
12 Thompson Rd.
East Windsor, CT 06088
Tel: 208/623-9901
Fax: 208/623-4657

Genie Products Inc.
Hwy. 64 East
P.O. Box 885
Rosman, NC 28772
Tel: 704/862-4772
Fax: 704/877-3480

Guyson Corp. of USA
W. J. Grande Industrial Pk.
Saratoga Springs, NY 12866-9044
Tel: 800/633-6677
Fax: 518/587-7840

H

Hardface Alloys Inc.
8351 Secura Way
Santa Fe Springs, CA 90670
Tel: 310/945-5477
Fax: 310/696-1868

Heany Industries Inc.
294 Briarwood Ln.
Scottsville, NY 14546
Tel: 716/889-2700
Fax: 716/889-2708

H.C. Starck Inc.
45 Industrial Pl.
Newton, MA 02161
Tel: 617/630-5920
Fax: 617/630-5919

Heream Enterprises
P.O. Box 861
8 Hyder St.
Westborough, MA 01581
Tel: 508/366-4923
Fax: 508/366-4923

HFW Industries
196 Philadelphia St.
Buffalo, NY 14027-0008
Tel: 716/875-3380
Fax: 716/875-3385

High Purity Gas Co.
4344 S. Main
Pearland, TX 77581-6043
Tel: 713/482-7002
Fax: 800/235-9380

Hitemco
160 Sweet Hollow Rd.
Old Bethpage, NY 11758
Tel: 516/752-7882
Fax: 516/752-7951

Hobart Tafa Technologies Inc.
146 Pembroke Rd.
Concord, NH 03301
Tel: 603/224-9585
Fax: 603/225-4342

Hoeganaes Corp.
River Road and Taylors Lane
Riverton, NJ 08053
Tel: 609/829-2220
Fax: 609/786-2574

I

Industrial Acoustics Co.
1160 Commerce Ave.
Bronx, Ny 10462
Tel: 718/931-8000
Fax: 718//863-1138

Inter American Export Co.—Div. of Cab New York Inc.
5 Oakland Pl.
Great Neck, NY 11021
Tel: 516/ 487-2288
Fax: 516/487-5408

I.S.P.A.
2915 Wilmarco Ave.
Baltimore, MD 21223
Tel: 410/644-4500
Fax: 410/644-1766

J

Jet-Com Inc.
201 Ballardvale St.
Wilmington, MA 01877
Tel: 617/933-8181
Fax: 508/658-0572

Jet Edge Inc.
825 Rhode Island Ave. S
Minneapolis, MN 55426
tel: 612/545-1477
Tel: 800/Jet-Edge
Fax: 612/545-5670

K

Kirk Industrial Sales Inc.
5168 Commercial Dr.
Yorkville, NY 13495
Tel: 315/736-7470
Fax: 315/736-1901

L

Leco Corp.
3000 Lakeview Ave.
St. Joseph, MI 49085-2396
Tel: 616/983-5531
Fax: 616/983-3850

Liquid Air Corp.
2121 N. California Blvd.
Walnut Creek, CA 94596
Tel: 510/977-6500
Fax: 510/977-6591

M

Machinists Inc.
751 S. Michigan St.
Seattle, WA 98108
Tel: 206/763-0990
Fax: 206/763-8709

Metallic Bonds Ltd.
50 Portland Ave.
P.O. Box 267
Beloit, WI 53512-0267
Tel: 608/365-9299
Fax: 608/365-2264

Metallizing Equipment Co. PVT Ltd.
5th Chopasni Rd.
Jodhpur Raj. 342 003
India
Tel: 91-291-35474
Fax: 91-291-35476

Metallurgical Technologies Inc.
14435 Max Rd.
Pearland TX 77581
Tel: 713/485-7765
Fax: 713/485-0211

Metco—Div. of Perkin-Elmer
1101 Prospect Ave.
P.O. Box 1006
Westbury, NY 11590-0201
Tel: 516/334-1300
Fax: 516/338-2123

Micron Metals Inc.
7186 W. Gates Ave.
Salt Lake City, UT 84120
Tel: 801/250-5919
Fax: 801/250-7295

Miller Thermal Inc.
555 Communication Dr.
Appleton, WI 54912
Tel: 414/731-6884
Fax: 414/734-2160

Montreal Carbide Co. Ltd.
1500 E. Nobel
Boucherville, Quebec J4B 5H3
Canada
Tel: 514/449-0147
Fax: 514/449-6101

Muscle Shoals Minerals Inc.
510 Mulberry Ln.
Cherokee, AL 25616
Tel: 205/370-7106
Fax: 205/370-7463

N

National Coating Technologies Inc.
1975 Logan Ave.
Winnepeg, Manitoba R2R 0H8
Canada
Tel: 204/632-5585
Fax: 204/694-3282

Norton Co.
1 New Bond St.
Worcester, MA 01606
Tel: 508/795-2351
Fax: 508/795-5868

O

Osram Sylvania Inc.
Hawes St.
Towanda, PA 18848
Tel: 717/265-2121
Fax: 717/265-1435

P

Performance Alloys Inc.
P.O. Box 670354
Dallas, TX 73367-0354
Tel: 214/350-0101
Fax: 214/350-0276

Plasma Coating Corp.
15331 S. Avalon Blvd.
Gardena, CA 90248
Tel: 310/532-3064
Fax: 310/532-0429

Plasmatec Inc.
1093, de Belleme
Boucherville, Quebec J4B 5Z4
Canada
Tel: 514/641-4175
Fax: 514/641-4934

Plasma-Tec Inc.
7500 Clyde Park SW
Grand Rapids, MI 49509
Tel: 616/455-2593
Fax: 616/455-4760

Plasma-Technik AG
17426 Daimler St.
Irvine, CA 92714

Tel: 714/863-1834
Fax: 714/863-6906

Plasma Technology Inc.
70 Rye St.
South Windsor, CT 06074-1218
Tel: 203/282-0650
Fax: 203/528-2631

Plasma Technology Inc.
1754 Crenshaw Blvd.
Torrance, CA 90501
Tel: 310/320-3373
Fax: 310/533-1677

Plastic Flamecoat Systems
1613 Hwy. 3
League City, TX 77573
Tel: 713/332-8180
Fax: 713/554-7434

Powder Alloy Corp.
10881 Millington Ct.
Cincinnati, OH 45242
Tel: 513/984-4016
Fax: 513/984-4017

Powder Feed Dynamics Inc.
24700 Highpoint Rd.
Beechwood, OH 44122
Tel: 216/484-7474
Fax: 216/484-7475

Praxair Surface Technologies Inc.
1500 Polco St.
Indianapolis, IN 46224
Tel: 317/240-2400
Fax: 317/240-2426

Praxair Surface Technologies Inc., Specialty Powders
1555 Main St.
Indianapolis, In 46224
Tel: 317/240-2208
Fax: 317/240-2225

Princeton Gamma Tech Inc.
1200 State Rd.
Princeton, NJ 08540
Tel: 609/924-7310
Fax: 924-1729

Progressive Technologies
4201 Patterson Ave. S.E.
Grand Rapids, MI 49512
Tel: 616/957-0871
Fax: 616/957-3484

Purtech Coatings Co.
R.R. 7, Box 7072
E. Stroudsburg, PA 18301
Tel: 717/424-1669
Fax: 717/424-0397

R

Racine Flame Spray Inc.
1810 William St.
Racine, WI 53404
Tel: 414/632-1989
Fax: 414/632-0681

RDB Enterprises Inc.
5049 1/2 Edith NE
Albuquerque, NM 87107
Tel: 506/344-7691

S

Saint Gobain/Norton Co. Industrial Ceramics Corp.
1 New Bond St.
Worcester, MA 01615-0008
Tel: 508/795-2488
Fax: 508/795-5011

Sermatech Technical Services
155 S. Limerick Rd.
Limerick, PA 19463
Tel: 215/948-2339
Fax: 216/948-0311

Sermatech Technical Services
12505 Reed Rd.
Sugar Land, TX 77478-2876
Tel: 713/240-1444
Fax: 713/240-4380

Sermatech Technical Services
507 Industrial Way
Boynton Beach, FL 34426-3644
Tel: 407/582-6080
Fax: 407/528-6505

Sermatech Technical Services
500 Hayden Station Rd.
Windsor, CT 06095-1368
Tel: 203/683-0711
Fax: 203/683-1631

Sermatech Technical Services
High Holburn Rd.
Codnor gate Industria Est.
Ripley, Derbyshire DE5 3NW
United Kingdom

Sermatech Technical Services
24 Landry St.
Biddeford, ME 04005-4310
Tel: 207/282-3787
Fax: 207/282-2694

Sermatech Technical Services
18502 Laurel Park Rd.
Compton, CA 90220

Tel: 310/604-0018
Fax: 310/635-3569

Saint Gobain/Norton Co. Industrial Ceramics Corp.
1 New Bond St.
P.O. Box 15008
Worcester, MA 01615-0008
Tel: 508/795-2488
Fax: 508/795-5011

Sherritt Gordon Ltd. Specialty Materials Div.
Fort Saskatchewan, Alberta
T8L 2P2
Canada
Tel: 403/992-5026
Fax: 403/992-5030

Shiva Tech Inc.
6260 South Bay Rd.
Cicero, NY 13039
Tel: 315/699-5332
Fax: 315/699-9899

Snyder General Corp.
215 Central Ave.
Louisville, KY 40208
Tel: 502/637-0011
Fax: 502/637-0351

Southwest Aeroservice Inc.
1501 E. 4th Pl.
Tulsa, OK 74120
Tel: 918/592-1177
Fax: 918/592-1551

Spray Masters Inc.
1023 W. 55th St., Ste. 202
Countryside, IL 60525
Tel: 708/482-4084
Fax: 708/482-4939

Spray Tech Inc.
P.O. Box 641
Sandyhook, CT 06482
Tel: 203/426-5112
Fax: 203/426-5498

St. Louis Metallizing Co.
4123 Sarpy Ave.
St. Louis, MO 63110
Tel: 314/531-5253
Fax: 314/531-3723

Stellite Coatings
1201 Eisenhower Dr. N
Goshen, IN 46526
Tel: 800/235-9353
Fax: 219/534-3417

Stoody Co.
5557 Nashville Rd.
Bowling Green, KY 42101
Tel: 800/227-9333
Fax: 800/243-4322

Stork Cellramic Inc.
8399 N. 87th St.
Milwaukee, WI 53224
Tel: 414/357-0260
Fax: 414/357-0267

Struers Inc.
26100 First St.
Westlake, OH 44145
Tel: 216/871-0071
Fax: 216/871-8188

Sulzer Plasma Technik Inc.
1972 Meijer Dr.
Troy, MI 48084
Tel: 313/288-1200
Fax: 313/288-4162

Superior Arc Metallizing Co. Inc.
3350-D Halls Mill Rd.
Mobile, AL 36606
Tel: 205/473-8500
Fax: 205/473-8505

Supfina Machine Co. Inc.
33 Planway Bldg. 1
Warwick, RI 02886
Tel: 401/738-4010
Fax: 401/738-4266

Surtech Inc.
20 Tuttle Pl.
Middleton, CT 06457
Tel: 203/632-9977
Fax: 203/635-2446

T

Technetics Corp.
1600 Industrial Dr.
Delano, FL 32724
Tel: 904/736-7373
Fax: 904/738-4533

Technical Laminations & Coatings
7125 W. Gunnison St.
Harwood Heights, IL 60556
Tel: 800/852-8273
Fax: 708/867-5654

Teledyne Hastings-Raydist
804 Newcombe Ave.
Hampton, VA 23669
Tel: 804/723-6531
Fax: 804/723-3925

Tekna Plasma Systems Inc.
3535 Industrial Blvd.
Sherbrooke, Quebec J1L 1X7
Canada
Tel: 819/820-2204
Fax: 819/820-2204

The Thermal Spray Laboratory Dept. of Metals Science &

Engineering S.U.N.Y. at Stony Brook
Stony Brook, NY 11794-2275
Tel: 516/632-8480
Fax: 516/632-8052

Thermal Coatings International Inc.
8305 S.E. Monterey #210
Portland, OR 97266
Tel: 503/654-5646
Fax: 503/654-5852

Thermal Spray Technologies Inc.
201 Frederick St.
Watertown, WI 53094
Tel: 414/261-0131
Fax: 414/261-4549

Thermion Metalizing Systems
P.O. Box 2136
Silverdale, WA 93838
Tel: 206/692-6656
Fax: 206/698-1539

Torit-Donaldson Co. Inc.
1400 W. 94th St.
Minneapolis, MN 55431
Tel: 612/887-3900
Fax: 612/887-3377

Tylan General
9577 Chesapeake Dr.
San Diego, CA 92123
Tel: 619/571-1222
Fax: 619/576-1703

U

UTP Welding Technology
10401 Greenbough
Houston, TX 77471
Tel: 713/499-1212
Fax: 713/499-4347

V

Valimet Inc.
431 E. Sperry Rd.
Stockton, CA 95206
Tel: 209/982-4870
Fax: 209/982-1365

Via Technologies
P.O. Box 2868
Fairfield, CA 94533-0286
Tel: 707/425-0365
Fax: 707/425-0365

W

Wall Colmonoy Corp.
30261 Stephenson Hwy.
Madison Heights, MI 48071
Tel: 313/585-6400
Fax: 313/585-7960

WHITCO
HCR 65 Box 77
South Strafford, VT 05070
Tel: 802/765-4104
Fax: 802/765-4104

Wynn Environmental Sales Co.
129 Terwood Road, Ste. 3
Willow Grove, PA 19090
Tel: 215/659-9667

Z

Zinco—Div. of La Salle Rolling Mills Inc.
1375 Ninth St.
La Salle, IL 61301
Tel: 815/224-3324
Fax: 815/224-4819

Zircoa Inc.
31501 Solon Rd.
Solon, OH 44139
Tel: 216/248-0500
fax: 216/248-8864

Supplemental Company Directory

Alcan Wire & Cable
Box 269
Toronto Dominion Centre
Toronto, Ontario M5K 1K1
Canada
(416) 860-4800
Duralcan[da] aluminum-aluminum oxide composite wire

Applied Powder Technologies
1415, 540 - 5th Ave., SW
Calgary, Alberta T2P 0M2
Canada
(403) 265-6700
novel oxide powder feedstocks

CEAC (Connecticut Engineering Associates Corp.)
P.O. Box 656
Sandy Hook, CT 06482
(203) 426-4733
powder feedstocks

Copperlok Inc.
5 Sunset Ln.
Hatboro, PA 19040
(215) 674-1661
antifouling coatings

Intermet Technology
116 Long Rd.
Georgetown, PA 15043
(412) 573-7754
innovative feedstock development, consulting

Montreal Carbide Co. LTD
1500-E Nobel

Boucherville, Quebec J4B 5H3
Canada
(514) 449-0147
boride and ceramic powder feedstocks

Muscle Shoals Minerals Co.
P.O. Box 518
Tuscumbia, AL 35674
(205) 370-7102
ceramic powder feedstocks

National Coatings
541 5th Ave.
McKeesport, PA 15132
(412) 678-6575
plastic thermal spray technology

Platt Bros. & Co.
670 S. Main St.
Waterbury, CT 0671
(203) 753-4194
zinc-based wire feedstocks

Specialty Powders
1555 Main St.
Indianapolis, IN 46224
(317) 240-2225
powder feedstocks

Surface Alloying & Wearfacing Services
4401 S. Pinemont, Ste. 200
Houston, Texas 77041
(713) 462-8030
innovative materials development, consulting

Wear Management Services
205 Station St.
Aliquippa, PA 15001
(412) 378-3535
wear-resistant feedstocks, consulting

Xform Inc.
100 N. Mohawk St.
Cohoes, NY 12047
(518) 233-7249
intermetallic powders

EQUIPMENT, SUPPLIES, AND SERVICES DIRECTORY

Directory Contents

THERMAL SPRAY EQUIPMENT

Combustion

A-Flame Corp.
A & A Co. Inc.
ASB Industries Inc.
Advanced Materials Technologies Inc.
Alamo Supply Co.
American Metallizing & Sandblasting Inc.
C & P Surface Technology
Cincinnati Thermal Spray-South
Douglas Call Co. Inc.
Eutectic Corp.
Exline Inc.
Flame Spray Engineering & Research
Flame Spray Northwest
General Plasma Inc.
Genie Products Inc.
Hardface Alloys Inc.
Hercam Enterprises
Hitemco
Inter American Export Co.—Div. of Cab New York, Inc.
Metallic Bonds Ltd.
Metallizing Equipment Co. PVT Ltd.
Metco—Div. of Perkin-Elmer
Miller Thermal Inc.
Plasmatec Inc.
Plastic Flamecoat Systems
Powder Feed Dynamics
Purtech Coatings Co.
RDB Enterprises Inc.
Surtech Inc.
UTP Welding Technology
Wall Colomonoy Corp.

High-Velocity Combustion

A & A Co. Inc.
ASB Industries Inc.
Advanced Materials Technologies Inc.
Air Products & Chemicals Inc.
Alamo Supply Co.
Alloy Carbide Co.
Cincinnati Thermal Spray-South
Douglas Call Co. Inc.
Drexel University
Electro-Plasma Inc.
Exline Inc.
Flame Spray Northwest
General Plasma Inc.
Hardface Alloys Inc.
Hercam Enterprises
Hitemco
Hobart Tafa Technologies Inc.
Inter American Export Co.—Div. of Cab New York Inc.
Metallic bonds Ltd.
Metallizing Equipment Co.
Metco—Div. of Perkin-Elmer

Miller Thermal Inc.
Montreal Carbide Co. Ltd.
Performance Alloys Inc.
Plasmatec Inc.
Plasma-Technik AG
Powder Feed Dynamics
Purtech Coatings Co.
Stellite Coatings
Sulzer Surface Tech
Surtech Inc.
The Thermal Spray Laboratory
UTP Welding Technology
WHITCO

Plasma

A-Flame Corp.
A & A Co. Inc.
ASB Industries Inc.
Advanced Materials Technologies Inc.
Air Products & Chemicals Inc.
Alamo Supply Co.
American Metallizing & Sandblasting Inc.
Applied Polymer Systems Inc.
Bay State Abrasives—Div. of Abrasive Industries Inc.
Cincinnati Thermal Spray-South
Douglas Call Co. Inc.
Drexel University
Electro-Plasma Inc.
Eutectic Corp.
Exline Inc.
Flame Spray Engineering & Research
Flame-Spray Industries Inc.
Flame-Spray Northwest
General Plasma Inc.
Genie Products Inc.
Hardface Alloys Inc.
Hereann Enterprises-Used Equipment Sales
Hitemco
Hobart Tafa Technologies Inc.
The ISPA Co.
Inter American Export Co.—Div. of Cab New York, Inc.
Metallic Bonds Ltd.
Metallizing Equipment Co. PVT Ltd.
Metco—Div. of Perkin-Elmer
Miller Thermal Inc.
Performance Alloys Inc.
Plasmatec Inc.
Powder Feed Dynamics Inc.
Purtech Coatings Co.
Sulzer Surface Tech
Surtech Inc.
Tekna Plasma Systems Inc.
The Thermal Spray Laboratory
WHITCO

Ceramic-Rod Combustion

Alamo Supply Company Inc.
ASB Industries Inc.
General Plasma Inc.
Hardface Alloys Inc.
Hereann Enterprises
Inter American Export Co.—Div. of Cab New York, Inc.
Metallic Bonds Ltd.
Metallizing Equipment Co. PVT Ltd.
Miller Thermal Inc.
Norton Co.
RDB Enterprises Inc.
Saint Gobain/Norton Co. Industrial Ceramics Corp.
Surtech Inc.
The Thermal Spray Laboratory

Wire Arc Spray

A-Flame Corp.
A & A Co. Inc.
Air Products & Chemicals Inc.
Alamo Supply Co.
American Metallizing & Sandblasting Inc.
Brodueur Machine Co. Inc.
C & P Surface Technology
Corpus Christi Coating & Machine Inc.
Douglas Call Co. Inc.
ECS Metals Ltd.
Exline Inc.
Flame Spray Engineering & Research
Flame-Spray Industries Inc.
Flame Spray Northwest
General Plasma Inc.
Genie Products Inc.
Hardface Alloys Inc.
Hereann Enterprises
Hitemco
Hobart Tafa Technologies Inc.
Inter American Export Co.—Div. of Cab New York, Inc.
The ISPA Co.
Metallic Bonds Ltd.
Metallizing Equipment Co. PVT Ltd.
Metco—Div. of Perkin-Elmer
Miller Thermal Inc.
Plasmatec International Inc.
Purtech Coatings Co.
RDB Enterprises Inc.
Superior Arc Metalizing Co. Inc.
Surtech Inc.
The Thermal Spray Laboratory
Thermion Metalizing Systems
WHITCO

PREPARATION EQUIPMENT AND SUPPLIES

Blasting Media
Bay State Abrasives—Div. of Abrasives Industries Inc.

Blast-It-All—Div. of Larry Hess & Assoc. Inc.
Cincinnati Thermal Spray-South
Douglas Call Co. Inc.
Hardface Alloys Inc.
Inter American Export Co.—Div. of Cab New York, Inc.
The ISPA Co.
Metallizing Equipment Co. PVT Ltd.
Metco—Div. of Perkin-Elmer
Montreal Carbide Co. Ltd.
Norton Co.
RDB Enterprises Inc.
Saint Gobain/Norton Co. Industrial Ceramics Corp.
The Thermal Spray Laboratory
Wall Colmonoy Corp.

Masking Compounds and Tapes

A-Flame Corp.
Alamo Supply Co.
Allied Signal/Fluorglas
Cincinnati Thermal Spray-South
Eutectic Corp.
Furon—CHR Division
Hardface Alloys Inc.
Hobart Tafa Technologies Inc.
Inter American Export Co.—Div. of Cab New York, Inc.
The ISPA Co,
Metallic Bonds Ltd.
Metallizing Equipment Co. PVT Ltd.
Metco—Div. of Perkin-Elmer
Performance Alloys Inc.
Plasmatec Inc.
Plasma Technik Inc.
Surtech Inc.
Technical Laminations & Coatings
UTP Welding Technology

Pressure and Vacuum Blasting

Alamo Supply Co.
Blast-It-All—Div. of Larry Hess & Assoc. Inc.
Cincinnati Thermal Spray-South
Empire Abrasive Equipment Co.
Guyson Corp. of USA
Hardface Alloys Inc.
Heream Enterprises
Inter American Export Co.—Div. of Cab New York, Inc.
Metallizing Equipment Co. PVT. Ltd.
Metco—Div. of Perkin-Elmer
Progressive Technologies
Surtech Inc.

THERMAL SPRAY PROCESS SUPPLIES

Carbide Powders

A-Flame Corp.
Alamo Supply Co.
Bay State Abrasives—Div. of Abrasive Industries Inc.
F.J. Brodmann & Co.
Cerac Inc.
Cincinnati Thermal Spray-South
Engineered Powder Technology
Eutectic Corp.
Inter American Export Co.—Div. of Cab New York, Inc.
Hardface Alloys Inc.
Heream Enterprises
H.C. Starck Inc.
Hobart Tafa Technologies Inc.
Hoeganaes Corp.
Inter American Export Co.—Div. of Cab New York, Inc.
Metallic Bonds Ltd.
Metallizing Equipment Co. PVT Ltd.
Metallurgical Technologies Inc.
Metco—Div. of Perkin-Elmer
Miller Thermal Inc.
Montreal Carbide Co. Ltd.
Osram Sylvania Inc.
Performance Alloys Inc.
Plasmatec Inc.
Powder Alloy Corp.
Praxair Surface Technologies Inc.
Specialty Powders
Sherritt Gordon Ltd.
Specialty Materials Div.
Spray Tech Inc.
Stellite Coatings
Plasma Technik Inc.
Surtech inc.
The Thermal Spray Laboratory
UTP Welding Technology
WHITCO

Ceramic Powders

A-Flame Corp.
Air Products & Chemicals
Alamo Supply Co.
A.O. Smith Corp.—Protective Coatings Div.
Bay State Abrasives—Div. of Abrasive Industries Inc.
F.J. Brodmann & Co.
Cerac Inc.
Cincinnati Thermal Spray-South
Engineered Powder Technology
Eutectic Corp.
Hardface Alloys Inc.
Heream Enterprises
H.C. Starck Inc.
Hobart Tafa Technologies Inc.
Inter American Export Co.—Div. of Cab New York, Inc.
Metallic Bonds Inc.
Metallizing Equipment Co. PVT Ltd.

Metallurgical
Technologies Inc.
Metco—Div. of Perkin-Elmer
Miller Thermal Inc.
Montreal Carbide Co. Ltd.
Muscle Shoals Minerals Inc.
Norton Co.
Osram Sylvania Inc.
Performance Alloys Inc.
Plasmatec Inc.
Powder Alloy Co.
Praxair Surface
Technologies Inc.
Specialty Powders
Saint Gobain/Norton Co.
Industrial Ceramics Corp.
Sulzer Plasma Technik Inc.
Surtech Inc.
The Thermal Spray
Laboratory
UTP Welding Technology
Zircoa Inc.

Ceramic Rods

Alamo Supply Co.
Cerac Inc.
F.J. Brodmann & Co.
Hardface Alloys Inc.
Heream Enterprises
Inter American Export Co.—
Div. Cab New York, Inc.
Metallic Bonds Inc.
Metallizing Equipment Co.
PVT Ltd.
Miller Thermal Inc.
Norton Co.
Performance Alloys Inc.
Saint Gobain/Norton Co.
Industrial Ceramics Corp.
Surtech Inc.
The Thermal Spray
Laboratory

Industrial Gases

Air Products & Chemical
Inc.
Cincinnati Thermal Spray-
South
Liquid Air Corp.

Metallizing Equipment Co.
PVT Ltd.
Spray Tech Inc.
The Thermal Spray
Laboratory

Metallic Powders

A-Flame Corp.
Air Products &
Chemicals Inc.
Alamo Supply Co.
Ametek Inc.
Amorphous Metals
Technologies Inc.
Anval
Applied Powder Technolo-
gies Inc.
Bay State Abrasives—Div.
of Abrasive Industries Inc.
Belmont Metals Inc.
F.J. Brodmann & Co.
Cerac Inc.
Cincinnati Thermal Spray-
South
Climax Specialty Metals
Dynamet Powder Products
Engineered Powder
Technology
Eutectic Corp.
Hardface Alloys Inc.
H.C. Starck Inc.
Heream Enterprises
Hobart Tafa
Technologies Inc.
Hoeganaes Corp.
Inter American Export Co.—
Div. of Cab New York, Inc.
Metallic Bonds Ltd.
Metallizing Equipment Co.
PVT. Ltd.
Metallurgical
Technologies Inc.
Metco—Div. of Perkin-Elmer
Micron Metals
Miller Thermal Inc.
Montreal Carbide Co.
Osram Sylvania Inc.
Performance Alloys Inc.
Plasmatec Inc.
Powder Alloy Corp.

Praxair Surface
Technologies Inc.
Specialty Powders
Sherritt Gordon Ltd.—
Specialty Mtls. Div.
Spray Tech Inc.
Stellite Coatings Stoody Co.
Stoody Co.
Sulzer Surface Tech
Surtech Inc.
The Thermal Spray
Laboratory
UTP Welding Technology
Valimet Inc.
Wall Colomonoy Corp.
WHITCO

Self-Fluxing Powders

A-Flame Corp.
Alamo Supply Co.
Ametek Inc.
Amorphous Metals
Technologies Inc.
Anval
Bay State Abrasives Div.
Abrasives Industries Inc.
F.J. Brodmann & Co.
Cincinnati Thermal Spray-
South
Dynamet Powder Products
Engineered Powder
Technology
Eutectic Corp.
Hardface Alloys Inc.
H.C. Starck Inc.
Hoeganaes Corp.
Inter American Export Co.—
Div. of Cab New York, Inc.
Metallic Bonds Ltd.
Metallizing Equipment Co
PVT Ltd.
Metallurgical
Technologies Inc.
Metco—Div. of Perkin-
Elmer
Miller Thermal Inc.
Montreal Carbide Co. Ltd.
Performance Alloys Inc.
Plasmatec International Inc.
Powder Alloy Corp.

Praxair Surface
Technologies Inc.
Specialty Powders Inc.
Spray Tech Inc.
Stellite Coatings
Plasma Technik Inc.
Surtech Inc.
The Thermal Spray
Laboratory
UTP Welding Technology
Wall Colmonoy Corp.

Other Powders

A-Flame Corp.
Alamo Supply Co.
F.J. Brodmann & Co.
Bay State Abrasives Div.
Abrasive Industries Inc.
Cerac Inc.
Cincinnati Thermal Spray-
South
Dynamet Powder Products
Engineered Powder
Technology
Eutectic Corp.
Hardface Alloys Inc.
H.C. Starck Inc.
Heream Enterprises
Hoeganaes Corp.
Inter American Export Co.—
Div. of Cab New York, Inc.
Metallic Bonds Ltd.
Metallurgical
Technologies Inc.
Metco—Div. of Perkin-Elmer
Miller Thermal Inc.
Montreal Carbide Co. Ltd.
Osram Sylvania Inc.
Performance Alloys Inc.
Plasmatec Inc.
Plastic Flamecoat Systems
Powder Alloy Corp.
Praxair Surface Technologies
Inc.
Specialty Powders
Sherritt Gordon Ltd.—
Specialty Mtls. Div.
Spray Tech Inc.
Stellite Coatings
Sulzer Surface Tech.
Surtech Inc.
The Thermal Spray
Laboratory
UTP Welding Technology

Wires

A-Flame Corp.
Air Products &
Chemicals Inc.
Alamo Supply Co.
Amorphous Metals
Technologies Inc.
(AMTECH)
Bay State Abrasives—Div.
of Abrasive Industries Inc.
Belmont Metals Inc.
F.J. Brodmann & Co.
Cincinnati Thermal Spray-
South
Douglas Call Co. Inc.
ECS Metals Ltd.
Eutectic Corp.
Flame-Spray Industries Inc.
Hardface Alloys Inc.
Heream Enterprises
Hobart Tafa Technologies
Inc.
Inter American Export Co.—
Div. of Cab New York, Inc.
Metallic Bonds Ltd.
Metco—Div. of Perkin-Elmer
Miller Thermal Inc.
Performance Alloys Inc.
Plasmatec International Inc.
Stellite Coatings
Stoody Co.
Superior Arc
Metalizing Co. Inc.
Surtech Inc.
The Thermal Spray
Laboratory
Thermion Metalizing
Systems
Wall Colmonoy Corp.
Zinco

Intermetallic Powders

Alamo Supply Co.
Anval
F.J. Brodman & Co.
Cerac Inc.
Cincinnati Thermal Spray-
South
Dynamet Powder Products
Engineered Powder
Technology
Eutectic Corp.
Hardface Alloys Inc.
H.C. Starck Inc.
Heream Enterprises
Hoeganaes Corp.
Inter American Export Co.—
Div. of Cab New York, Inc.
Metallic Bonds Ltd.
Metallurgcal
Technologies Inc.
Montreal Carbide Co. Ltd.
Osram Sylvania Inc.
Performance Alloys Inc.
Praxair Surface
Technologies Inc.
Specialty Powders
Spray Tech Inc.
Stellite Coatings
Surtech Inc.
The Thermal Spray
Laboratory
Wall Colmonoy Corp.

Thermal Spray Masking Tapes

Technical Laminations &
Coatings

THERMAL SPRAY FINISHING SUPPLIES

Finishing Equipment

Alamo Supply Co.
Empire Abrasive
Equipment Co.

Guyson Corp. of USA
Hardface Alloys Inc.
Hercam Enterprises
Inter American Export Co.—Div. of Cab New York, Inc.
Plastic Flamecoat Systems
Ultra Precision Technologies Inc. (UPTEC)

Finishing Tools/ Grinding Wheels

Alamo Supply Co. Inc.
Bay State Abrasives—Div. of Abrasive Industries Inc.
Inter American Export Co.—Div. of Cab New York, Inc.
Norton Co.
Saint Gobain/Norton Co. Industrial Ceramics Corp.
Ultra Precision Technologies Inc. (UPTEC)

Grinders/Machine Tools

Alamo Supply Co.
Hercam Enterprises
Inter American Export Co.—Div. of Cab New York, Inc.
Metco—Div. of Perkin-Elmer
Supfina Machine Co. Inc.
Torit-Donaldson Co. Inc.

Sealants

A B Seals Inc.
Alamo Supply Co.
Cincinnati Thermal Spray-South
Eutectic Corp.
Hardface Alloys Inc.
Inter American Export Co.—Div. of Cab New York, Inc.
Metallic Bonds Ltd.
Metallizing Equipment Co.
Metco—Div. of Perkin-Elmer
Miller Thermal Inc.
Montreal Carbide Co. Ltd.
Performance Alloys Inc.
Surtech Inc.
UTP Welding Technology

FINISHING SERVICES

Machining

A & A Co. Inc.
ASB Industries Inc.
Advanced Materials Technologies Inc.
Aerobraze Corp.—Coatings
Alloy Carbide Co.
American Metallizing & Sandblasting Inc.
Bay State Abrasive—Div. of Abrasive Industries Inc.
B.C. Metals Inc.
Bender Machine Inc.
Bender Machine Services Ltd.
Brodeur Machine co. Inc.
Cincinnati Thermal Spray-South
Exline Inc.
Fisher Products Inc.
Flame Spray Coatings Co. Inc.
Flame Spray Engineering & Research
Flame Spray Northwest
F.W. Gartner Thermal Spraying Co.
General Magnaplate Corp.
General Plasma Inc.
HFW Industries
Hitemco
Jet-Com Inc.
Kirk Industrial Sales Inc.
Machinists Inc.
Metallic Bonds Ltd.
Performance Alloys Inc.
Plasma Coating Corp.
Plasma-Tec Inc.
Plasma Technology Inc.
Purtech Coatings Co.
Racine
RDB Enterprises Inc.
Sermatech Technical Services
Southwest Aeroservice Inc.
St. Louis Metallizing Co.
Supfina Machine Co. Inc.
Technetics Corp.
Thermal Coatings International Inc.

Grinding

A & A Co. Inc.
Advanced Materials Technologies Inc.
Alloy Carbide Co.
American Metallizing & Sandblasting Inc.
APS—Materials Inc.
B.C. Metals Inc.
Bender Metals Inc.
Bender Machine Inc.
Cincinnati Thermal Spray-South
Exline Inc.
Fisher Products Inc.
Flame Spray Coating Co. Inc.
Flamespray Engineering & Research
Flame Spray Northwest
F.W. Gartner Thermal Spraying Co.
General Plasma Inc.
HFW Industries Inc.
Hitemco
The ISPA Co.
Jet-Com Inc.
Machinists Inc.
Metallic Bonds Ltd.
Performance Alloys Inc.
Plasma Coating Corp.
Plasma-Tec Inc.
Plasma Technology Inc.
Racine
RDB Enterprises Inc.
Sermatech Technical Services
Technetics Corp.
Thermal Coatings Intl. Inc.

Superfinishing

A & A Co. Inc.

Alloy Carbide co.
American Metallizing & Sandblasting Inc.
Bender Machine Inc.
Cincinnati Thermal Spary-South
Exline Inc.
Flame Spray Coating Co. Inc.
Flamespray Engineering & Research
Flamespray Northwest
F.W. Gartner Thermal Spraying Co.
Hardface Alloys Inc.
HFW Industries Inc.
Hitemco
Inter American Export Co.—Div. of Cab New York Inc.
Machinists Inc.
Metallic Bonds Ltd.
Performance Alloys Inc.
Plasma Coating Corp.
Plasma Technology Inc.
Sermatech Technical Services
Thermal Coatings Intl. Inc.
Ultra Precision Technologies Inc. (UPTEC)

Other

A & A Co. Inc.
Alloy Carbide Co.
ASB Industries Inc.
Advanced Materials Technologies Inc.
Alloy Carbide Co.
American Metallizing & Sandblasting Inc.
Bender Machine Inc.
Bender Machine Services Ltd.
Brodueur Machine Co. Inc.
C & P Surface Technology
Exline Inc.
Flame Spray Northwest
F.W. Gartner Thermal Spraying Co.
General Magnaplate Corp.
General Plasma Inc.
HFW Ind.
Hitemco
The ISPA Co.
Machinists Inc.
Metallic Bonds Ltd.
Plasma Coating Corp.
Racine
Sermatech Technical Services
Southwest Aeroservice Inc.
St. Louis Metallizing Co.
Supfina Machine Co. Inc.
Thermal Coatings Intl. Inc.

AUXILIARY THERMAL SPRAY EQUIPMENT

Air Compressors

Alamo Supply Co. Inc.
Alloy Carbide Co.
Cincinati Thermal Spray-South
Douglas Call Co. Inc.
Inter American Export Co.—Div. of Cab New York, Inc.
The Thermal Spray Laboratory

Automated Spraying Systems

A-Flame Corp.
Alamo Suply Co. Inc.
Bay State Abrasive—Div. of Abrasive Industries Inc.
Control Vision Inc.
Electro-Plasma Inc.
Engineered Coatings Inc.
Flame-Spray Industries Inc.
Hardface Alloys Inc.
Heream Enterprises
Hobart Tafa Technologies Inc.
Inter American Export Co.—Div. of Cab New York, Inc.
Metallic Bonds Ltd.
Metallizing Equipment Co. PVT Ltd.
Metco—Div. of Perkin-Elmer
Miller Thermal Inc.
Plasmatec Inc.
Plasma-Technik AG
Plastic Flamecoat Systems
Progressive Technologies
Saint Gobain/Norton Co. Industrial Ceramics Corp.
Sulzer Surface Tech
Surtech Inc.
The Thermal Spray Laboratory

Exhaust Systems/ Baghouses

A-Flame Corp.
Alamo Supply Co.
Bay State Abrasives—Div. of Abrasive Industries Inc.
Blast-It-All—Div. of Larry Hess & Assoc. Inc.
Cincinnati Thermal Spray-South
DCE Inc.
Douglas Call Co. Inc.
Hardface Alloys Inc.
Hobart Tafa Technologies Inc.
Inter American Export Co.—Div. of Cab New York, Inc.
Metallic Bonds Ltd.
Metallizing Equipment Co. PVT Ltd.
Metco—Div. of Perkin-Elmer
Miller Thermal Inc.
Plasmatec International Inc.
Sulzer Surface Tech
Surtech Inc.
Torit-Donaldson Co. Inc.
UTP Welding Technology
Wynn Environmental Sales Co.

Flow Controller

A-Flame Corp.
Air Products & Chemicals Inc.
Bay State Abrasives—Div.

of Abrasive Industries Inc.
Cincinnati Thermal Spray-South
Engineered Coatings Inc.
Hardface Alloys Inc.
Hobart Tafa Technologies Inc.
Miller Thermal Inc.
Plasmatec Inc.
Teledyne Hastings-Raydist
Tylan General

Manipulators

A-Flame Corp.
Alamo Supply Co. Inc.
Cincinati Thermal Spray-South
Eutectic Corp.
Hardface Alloys Inc.
Heream Enterprises
Hobart Tafa Technologies Inc.
Inter American Export Co.—Div. of Cab New York, Inc.
Metallic Bonds Ltd.
Metallizing Equipment Co. PVT Ltd.
Metco—Div. of Perkin-Elmer
Miller Thermal Inc.
Plasmatec Inc.
Progressive Technologies
Sulzer Surface Tech
Surtech Inc.
The Thermal Spray Laboratory

Metallographic Equipment

Buehler Ltd.
Cincinnati Thermal Spray-South
Heream Enterprises
Leco Corp.
Princeton Gamma-Tech Inc.
The Thermal Spray Laboratory

Powder Feeder

A-Flame Corp.
Alamo Supply Co.
Bay State Abrasives—Div. of Abrasive Industries Inc.
Cincinnati Thermal Spray-South
Electro-Plasma Inc.
Eutectic Corp.
Hardface Alloys Inc.
Heream Enterprises
Hobart Tafa Technologies Inc.
Inter American Export Co.—Div. of Cab New York, Inc.
Metallic Bonds Ltd.
Metallizing Equipment Co. PVT. Ltd.
Metco—Div. of Perkin-Elmer
Miller Thermal Inc.
Plasmatec Inc.
Plasma-Technik AG
Stellite Coatings
Sulzer Surface Tech
Surtech Inc.
The Thermal Spray Laboratory
UTP Welding Technology

Robotics

A-Flame Corp.
Cincinnati Thermal Spray-South
Control Vision inc.
Eutectic Corp.
Inter American Export Co.—Div. of Cab New York, Inc.
Hardface Alloys Inc.
Heream Enterprises
Hobart Tafa Technologies Inc.
Metallic Bonds Ltd.
Metco—Div. of Perkin-Elmer
Miller Thermal Inc.
Plasmatec Inc.
Progressive Technologies
Sulzer Surface Tech
Surtech Inc.
The Thermal Spray Laboratory
Via Technologies

Soundproof Rooms

A-Flame Corp.
Alamo Supply Co.
Cincinnati Thermal Spray-South
Heream Enterprises
Hobert Tafa Technologies Inc.
Industrial Acoustics Co.
Inter American Export Co.—Div. of Cab New York, Inc.
Metallic Bonds Ltd.
Metallizing Equipment Co. PVT Ltd.
Metco—Div. of Perkin-Elmer
Miller Thermal Inc.
Plasmatec International Inc.
Sulzer Surface Tech
Surtech Inc.

Spray Booths

A-Flame Corp.
A B Seals Inc.
Alamo Supply Co.
Bay State Abrasive—Div. of Abrasive Industries Inc.
Blast-It-All—Div. of Larry Hess & Assoc. Inc.
Cincinnati Thermal Spray-South
Electro-Plasma Inc.
Hardface Alloys Inc.
Heream Enterprises
Hobart Tafa Technologies Inc.
Industrial Acoustics Inc.
Inter American Export Co.—Div. of Cab New York, Inc.
Metallic Bonds
Metallizing Equipment Co. PVT Ltd.
Metco—Div. of Perkin-Elmer

Miller Thermal inc.
Plasmatec Inc.
Plasma-Technik AG
Sulzer Surface Tech
Surtech Inc.
The Thermal Spray Laboratory
Wynn Environmental Sales Co.

Testing Equipment

Buehler Ltd.
Cincinnati Thermal Spray-South
Control Vision Inc.
Engineered Coatings Inc.
Heream Enterprises
Princeton Gamm-Tech Inc.
Surtech Inc.
The Thermal Spray Laboratory

Wet Collectors

A-Flame Corp.
Alamo Supply Co.
Bay State Abrasives—Div. of Abrasive Industries Inc.
Cincinnati Thermal Spray-South
Hardface Alloys Inc.
Hobart Tafa Technologies Inc.
Inter American Export Co.—Div. of Cab New York, Inc.
Metallizing Equipment Co. PVT. Ltd.
Metco—Div. of Perkin-Elmer
Miller Thermal Inc.
Snyder General Corp.
Sulzer Surface Tech
Surtech Inc.
The Thermal Spray Laboratory
Wynn Environmental Sales Co.

TESTING, R&D, EDUCATION

Contract Research

A & A Co. Inc.
Adiabatics Inc.
Aerobraze Corp.—Coatings
Air Products & Chemicals Inc.
Applied Polymer Systems Inc.
APS—Materials Inc.
F.J. Brodmann & Co.
Cincinnati Thermal Spray Inc.
Cincinnati Thermal Spray-South
Drexel University
Electro-Plasma Inc.
Ellison Surface Technologies Inc.
Engineered Coatings Inc.
Engineered Powder Technology
Flame Spray Coating Co. Inc.
Flame-Spray Industries Inc.
General Magnaplate Inc.
General Plasma Inc.
Hardface Alloys Inc.
Heany Industries Inc.
Hobart Tafa Technologies Inc.
Hitemco
The ISPA Co.
Metallic Bonds Ltd.
Metallizing Equipment Co. PVT Ltd.
Miller Thermal Inc.
Norton Co.
Plasma Coating Corp.
Plasma-Technik AG
Plasma Technology Inc.
Plasma Technology Inc.
Sherrit Gordon Ltd.—Specialty Mtls. Div.
Sulzer Surface Tech.
Technetics Corp.
Tekna Plasma Systems Inc.
The Thermal Spray Laboratory
Thermal Spray Technologies Inc.
Via Technologies
WHITCO

Market Research

Adiabatics Inc.
Blakely & Associates
F.J. Brodmann & Co.
Cincinnati Thermal Spray-South
Drexel University
General Plasma Inc.
Metallic Bonds Ltd.
The Thermal Spray Laboratory
Via Technologies

Education

A-Flame Corp.
Advanced Materials Technologies Inc.
Cincinnati Thermal Spray-South
Drexel University
Electro-Plasma Inc.
Eutectic Corp.
Hardface Alloys Inc.
Hitemco
Hobert Tafa Technologies Inc.
Metallizing Equipment Co. PVT. Ltd.
Montreal Carbide Co. Ltd.
Sulzer Surface Tech
Tekna Plasma Systems Inc.
The Thermal Spray Laboratory
Via Technologies
Wall Colmonoy Corp.

Testing Services/ Equipment/Supplies

A & A Co. Inc.
B.C. Metals Inc.

Buehler Ltd.
Cincinnati Thermal Spray Inc.
Cincinnati Thermal Spray-South
Control Vision Inc.
Drexel University
Electro-Plasma Inc.
Ellison Surface Technologies Inc.
Engineered Coatings Inc.
Flame Spray Coating Co. Inc.
General Plasma Inc.
Hitemco
Metallurgical Equipment Co. PVT. Ltd.
Miller Thermal Inc.
Norton Co.
Plasma-Technik AG
Princeton Gamma-Tech Inc.
Shiva Tech Inc.
Struers Inc.
Sulzer Surface Tech
Supfina Machine Co. Inc.
Tekna Plasma Systems Inc.
The Thermal Spray Laboratoy
Thermal Spray Technologies Inc.
Wall Colmonoy Corp.

Microscopy

Aerobraze Corp.—Coatings
Air Products & Chemicals Inc.
Amorphous Metals Technologies Inc. (AMTECH)
F.J. Brodmann & Co.
Buehler Ltd.
Cincinnati Thermal Spray-South
Drexel University
Ellison Surface Technologies Inc.
Flame Spray Inc.
General Plasma Inc.
Heany Industries Inc.
Hitemco
Miller Thermal Inc.
Montreal Carbide Co. Ltd.
National Coating Technologies Inc.
Norton Co.
Princeton Gamma-Tech Inc.
The Thermal Spray Laboratory
Thermal Spray Technologies Inc.

Test Coupons

A-Flame Corp.
A & A Co. Inc.
Adiabatics Inc.
Advanced Materials Technologies Inc.
Air Products & Chemicals Inc.
B.C. Metals Inc.
Cincinnati Thermal Spray-South
Douglas Call Co. Inc.
Drexel University
Ellison Surface Technologies Inc.
Engineered Coatings Inc.
Eutectic Corp.
Flame Spray Coating Co. Inc.
General Plasma Inc.
Heany Industries Inc.
Hobart Tafa Technologies Inc.
The ISPA Co.
Metallizing Equipment Co. PVT Ltd.
Miller Thermal Inc.
National Coating Technologies Inc.
Performance Alloys Inc.
Plasma Coating Corp.
Plasma Technology Inc.
Plasma Technology Inc.
The Thermal Spray Laboratory
Thermal Spray Technologies Inc.

THERMAL SPRAY APPLICATORS

On-Site Coating Services

A & A co. Inc.
Adiabatics Inc.
Aerobraze Corp..—Coatings
Alamo supply Co.
Alloy Carbide Co.
American Metallizing & Sandblasting Inc.
Applied Polymer Systems Inc.
APS-Materials Inc.
ASB Industries Inc.
Bender Machine Inc.
Bender Machine Services Ltd.
Brodueur Machine Co. Inc.
Cincinnati Thermal Spray Inc.
Cincinnati Thermal Spray-South
Douglas Call Co. Inc.
Ellison Surface Technologies Inc
Engineered Coatings Inc.
Flame Spray Coating Co. Inc.
Flame Spray Engineering & Research
Flame Spray Northwest
F.W. Gartner Thermal Spraying Co.
The ISPA Co.
Metallic Bonds Ltd.
National Coating Technologies Inc.
Plasma-Technik AG
Purtech Coatings Co.
Spray Masters Inc.
Sulzer Escher Wyss
Superior Arc Metalizing Co. Inc.
Sermatech Technical Services
Thermal Coatings

International Inc.

Shop Coating Services

A & A Co. Inc.
Adiabatics Inc.
Advanced Materials Technologies Inc.
Aeroberaze Corp.—Coatings
Alamo Supply Co.
Alloy Carbide Co.
American Metallizing & Sandblasting Inc.
A.O. Smith Corp.—Protective Coatings Div.
Applied Polymer Systems Inc.
APS-Materials Inc.
ASB Industries Inc.
Bender Machine Inc.
Bender Machine Services Ltd.
Brodeur Machine Co. Inc.
C & P Surface Technology
Cincinnati Thermal Spray Inc.
Cincinnati Thermal Spray—South
Douglas Call Co. Inc.
Electro-Plasma Inc.
Ellison Surface Technologies Inc.
Engineered Coatings Inc.
Fisher Products Inc.
Flame Spray Engineering & Research
Flame Spray Coating Co. Inc.
Flame Spray Northwest
F.W. Gartner Thermal Spraying Co.
General Magnaplate Corp.
General Plasma Inc.
Genie Products Inc.
Heany Industries Inc.
HFW Industries
Hitemco
The ISPA Co.
Jet-Com Inc.
Kirk Industrial Sales Inc.
Metallic Bonds Ltd.
Metallizing Equipment Co. PVT Ltd.
National Coating Technologies Inc.
Plasma Coating Corp.
Plasma-Tec Inc.
Plasma Technology Inc.
Plasma Technology Inc.
Purtech Coatings Co.
RDB Enterprise Inc.
Southwest Aeroservise Inc.
Spray Masters Inc.
St. Louis Metallizing Co.
Stork Cellramic Inc.
Technetics Corp.
Thermal Coatings International Inc.
Thermal Spray Technologies Inc.
Union Carbide Coatings Service Corp.
Wall Colmonoy Corp.

Volume Production

A & A Co. Inc.
Alamo supply Co. Inc.
Advanced Materials Technologies Inc.
Aerobraze Corp.—Coatings
Alloy Carbide Co.
American Metallizing & Sandblasting Inc.
A.O. Smith Corp.—Protective Coatings Div.
APS -Materials Inc.
ASB Industries Inc.
Bender Machine Inc.
Brodeur Machine Co. Inc.
Cincinnati Thermal Spray Inc.
Cincinnati Thermal Spray—South
Douglas Call Co. Inc.
Ellison Surface Technologies Inc.
Engineered Coatings Inc.
Fisher Products Inc.
Flame Spray Coating Co. Inc.
Flame Spray Engineering & Research
Flame Spray Co. Inc.
F.W. Gartner Thermal Spraying Co.
General Plasma Inc.
Heany Industries Inc.
Hitemco
Jet-Com Inc.
Kirk Industrial Sales Inc.
Metallic Bonds Ltd.
Metallizing Equipment Co. PVT Ltd.
National Coating Technologies Inc.
Plasma Coating Corp.
Plasma-Tec Inc.
Plasma Technology Inc.
Plasma Technology Inc.
Purtech Coatings Co.
Sermatech Technical Services
Southwest Aeroservice Inc.
Spray Masters Inc.
St. Louis Metallizing Co.
Stork Cellramic Inc.
Technetics Corp.
Thermal Coatings International Inc.
Thermal Spray Technologies Inc.

THERMAL SPRAY REMOVAL

Water and Cleaning Systems

General Plasma Inc.
Hobart Tafa Technologies Inc.
Jet Edge Inc.
Metallic Bonds Ltd.
Progressive Technologies

Other

Cincinnati Thermal Spray-South
Douglas Call Co. Inc.
Exline Inc.

Flame Spray
Coating Co. Inc.
Flamespray Engineering &
Research
General Plasma Inc.
Montreal Carbide Co. Ltd.

THERMAL SPRAY APPLICATIONS

Electrical/Electronics

A & A Co. Inc.
Adiabatics Inc.
Advanced Materials
Technologies Inc.
Aerobraze Corp.—Coatings
Air Products &
Chemicals Inc.
Alamo Supply Co.
American Metallizing &
Sandblasting Inc.
A.O. Smith Corp.—
Protective Coatings Div.
APS-Materials Inc.
ASB Industries Inc.
Bay State Abrasives—Div.
Abrasives Industries Inc.
Bender Machine Inc.
Bender Machine
Services Ltd.
Brodeur Machine Co. Inc.
Cincinnati Thermal
Spray Inc.
Cincinnati Thermal Spray—
South
Douglas Call Co. Inc.
Ellison Surface
Technologies Inc.
Engineered Coatings Inc.
Eutectic Corp.
Fisher Products Inc.
Flame Spray Engineering &
Research
Flame-Spray Industries Inc.
Flame Spray
Coating Co. Inc.
F.W. Gartner Thermal
Spraying Co.
General Magnaplate corp.
General Plasma Inc.
Heany Industries Inc.
Hitemco
Hobart Tafa
Technologies Inc.
Jet-Com Inc.
Kirk Industrial Sales Inc.
Metallic Bonds Ltd.
Metco—Div. of Perkin-Elmer
National Coating
Technologies Inc.
Plasma Coating Corp.
Plasma-Tec Inc.
Plasma Technology Inc.
Plastic Flamecoat Systems
Purtech Coatings Co.
RDB Enterprise Inc.
Southwest Aeroservise Inc.
St. Gobain/Norton Industrial
Cermaics Corp.
St. Louis Metallizing Co.
Stork Cellramic Inc.
Sulzer Plasma Technik Inc.
Surtech Inc.
The Thermal Spray
Laboratory
Thermal Spray
Technologies Inc.
Union Carbide Coatings
Service Corp.

Wear Resistance

A & A Co. Inc.
Adiabatics Inc.
Advanced Materials
Technologies Inc.
Aerobraze Corp.—Coatings
Air Products &
Chemicals Inc.
Alamo Supply Co.
Alloy Carbide Co.
American Metallizing &
Sandblasting Inc.
Amorphous Metals
Technologies Inc.
A.O. Smith Corp.—
Protective Coatings Div.
Applied Polymer
Systems Inc.
AS-Materials Inc
ASB Industries Inc.
Bay State Abrasives—Div.
Abrasive Industries Inc.
Bender Machine Inc.
Bender Machine
Services Ltd.
Brodeur Machine Co. Inc.
C & P Surface Technology
Cincinnati Thermal
Spray Inc.
Cincinnati Thermal
Spray—South
Douglas Call Co. Inc.
Drexel University
Ellison Surface
Technologies Inc.
Engineered Coatings Inc.
Eutectic Corp.
Fisher Products Inc.
Flame Spray
Coating Co. Inc.
Flame Spray Engineering &
Research
Flame-Spray Industries Inc.
Flame Spray
Coating Co. Inc.
Flame Spray Northwest
General Magnaplate corp.
General Plasma Inc.
Heany Industries Inc.
HFW Industries
Hitemco
Hobart Tafa
Technologies Inc.
The ISPA Co.
Jet-Com Inc.
Kirk Industrial Sales Inc.
Machinists Inc.
Metallic Bonds Ltd.
Metallizing Equipment Co.
PVT Ltd.
Metco—Div. of Perkin-Elmer
National Coating
Technologies Inc.
Plasma Coating Corp.
Plasma-Tec Inc.
Plasma Technology Inc.
Plasma Technology Inc.
Powder Alloy Corp.
Praxair Surface

Technologies Inc.
Purtech Coatings Co.
RDB Enterprise Inc.
Sermatech Technical Services
Southwest Aeroservise Inc.
Spray Masters Inc.
St. Gobain/Norton Industrial Ceramics Corp.
St. Louis Metallizing Co.
Stellite Coatings
Superior Arc Metalizing Co. Inc.
Surtech Inc.
Technetics Corp.
The Thermal Spray Laboratory
Thermal Coatings International Inc.
Thermal Spray Technologies Inc.
Wall Colmonoy Corp.

Reclamation

A & A Co. Inc.
Advanced Materials Technologies Inc.
Aerobraze Corp.—Coatings
Air Products & Chemicals Inc.
Alamo Supply Co.
Alloy Carbide Co.
American Metallizing & Sandblasting Inc.
APS-Materials Inc.
ASB Industries Inc.
Bender Machine Inc.
Bender Machine Services Ltd.
Brodeur Machine Co. Inc.
Cincinnati Thermal Spray Inc.
Cincinnati Thermal Spray—South
Ellison Surface Technologies Inc.
Engineered Coatings Inc.
Eutectic Corp.
Exline Inc.
Fisher Products Inc.
Flame Spray Engineering & Research
Flame-Spray Industries Inc.
Flame Spray Coating Co. Inc.
Flame Spray Northwest
F.W. Gartner Thermal Spraying Co.
General Magnaplate corp.
General Plasma Inc.
Heany Industries Inc.
HFW Industries
Hitemco
Hobart Tafa Technologies Inc.
Jet-Com Inc.
Kirk Industrial Sales Inc.
Machinists Inc.
Metallic Bonds Ltd.
Metco—Div. of Perkin-Elmer
National Coating Technologies Inc.
Plasma Coating Corp.
Plasma-Tec Inc.
Plasma Technology Inc.
Praxair Surface Technologies Inc.
Purtech Coatings Co.
RDB Enterprise Inc.
Sermatech Technical Services
Southwest Aeroservise Inc.
St. Gobain/Norton Industrial Ceramics Corp.
St. Louis Metallizing Co.
Stellite Coatings
Stork Cellramics Inc.
Sulzer Plasma Technik Inc.
Superior Arc Metalizing Co. Inc.
Surtech Inc.
The Thermal Spray Laboratory
Thermal Coatings International Inc.
Thermal Spray Technologies Inc.
Union Carbide Coatings Service Corp.
Wall Colmonoy Corp.

Clearance Control

A & A Co. Inc.
Advanced Materials Technologies Inc.
Aerobraze Corp.—Coatings
Alamo Supply Co. Inc.
ASB Industries Inc.
American Metallizing & Sandblasting Inc.
APS-Materials Inc.
Bender Machine Inc.
Bender Machine Services Ltd.
Cincinnati Thermal Spray Inc.
Cincinnati Thermal Spray—South
Douglas Call Co. Inc.
Ellison Surface Technologies Inc.
Engineered Coatings Inc.
Fisher Products Inc.
Flame Spray Engineering & Research
Flame Spray Coating Co. Inc.
Flame Spray Industries Inc.
Flame Spray Northwest
F.W. Gartner Thermal Spraying Co.
General Plasma Inc.
Heany Industries Inc.
HFW Industries
Hitemco
Hobart Tafa Technologies Inc.
Jet-Com Inc.
Kirk Industrial Sales Inc.
Machinists Inc.
Metallic Bonds Ltd.
Metallizing Equipment Co. PVT Ltd.
Metco—Div. of Perkin-Elmer
National Coating Technologies Inc.
Plasma Coating Corp.
Plasma Technology Inc.
Plasma Technology Inc.
Plasma-Tec Inc.
Praxair Surface

Technologies Inc.
Purtech Coatings
RDB Enterprises
Southwest Aeroservise Inc.
St. Gobain/Norton Industrial
Ceramics Corp.
St. Louis Metallizing Co.
Sulzer Plasme Technik Inc.
Surtech Inc.
Technetics Corp.
The Thermal Spray
Laboratoy
Thermal Coatings
International Inc.
Thermal Spray
Technologies Inc.

Thermal Barrier

A & A Co. Inc.
Adiabatics Inc.
Advanced Materials
Technologies Inc.
Aerobraze Corp—Coatings
Alamo Supply Co.
American Metallizing &
Sandblasting Inc.
Applied Polymer
Systems Inc.
APS-Materials Inc.
ASB Industries Inc.
Bay State Abrasives—Div.
Abrasive Industries Inc.
Bender Machine Inc.
Bender Machine
Services Ltd.
Brodeur Machine Co. Inc.
Cincinnati Thermal
Spray Inc.
Cincinnati Thermal Spray—
South
Douglas Call Co. Inc.
Ellison Surface
Technologies Inc.
Engineered Coatings Inc.
Eutectic Corp.
Fisher Products Inc.
Flame Spray Engineering &
Research
Flame Spray Coating Co. Inc
Flame-Spray Industries Inc.
Flame Spray Northwest
F.W. Gartner Thermal
Spraying Co.
General Magnaplate corp.
General Plasma Inc.
Heany Industries Inc.
HFW Industries
Hitemco
Hobart Tafa
Technologies Inc.
Jet-Com Inc.
Kirk Industrial Sales Inc.
Machinists Inc.
Metallic Bonds Ltd.
Metallizing Equipment Co.
PVT Ltd.
Metco—Div. of Perkin-Elmer
National Coating
Technologies Inc.
Plasma Coating Corp.
Plasma-Tec Inc.
Plasma Technology Inc.
Powder Alloy Corp.
Praxair Surface
Technologies Inc.
Purtech Coatings Co.
RDB Enterprise Inc.
Sermatech Technical
Services
Southwest Aeroservise Inc.
Spray Masters Inc.
St. Louis Metallizing Co.
Stork Cellramic Inc.
Sulzer Plasma Technik Inc.
Superior Arc
Metallizing Co. Inc.
Surtech Inc.
Technetics Corp.
The Thermal Spray
Laboratory
Thermal Coatings
International Inc.
Thermal Spray
Technologies Inc.
Wall Colmonoy Corp.

High Temperature Corrosion

A & A Co. Inc.
Adiabatics Inc.
Advanced Materials
Technologies Inc.
Aerobraze Corp—Coatings
Air Products &
Chemicals Inc.
Alamo Supply Co.
Alloy Carbide Co.
American Metallizing &
Sandblasting Inc.
A.O. Smith Corp.—
Protective Coatings Div.
Applied Polymer Systems
APS-Materials Inc.
ASB Industries Inc.
Bay State Abrasives—Div.
Abrasives Industries Inc.
Bender Machine Inc.
Bender Machine
Services Ltd.
Brodeur Machine Co. Inc.
Cincinnati Thermal
Spray Inc.
Cincinnati Thermal Spray—
South
Douglas Call Co. Inc.
Drexel University
Ellison Surface
Technologies Inc.
Engineered Coatings Inc.
Eutectic Inc.
Fisher Products Inc.
Flamespray Engineering &
Research
Flame Spray Coating Inc.
Flame-Spray Industries Inc
Flame Spray Northwest
F.W. Gartner Thermal
Spraying Co.
General Magnaplate Corp.
General Plasma Inc.
Heany Industries Inc.
HFW Industries Inc.
Hitemco
Hobart Tafa
Technologies Inc.
The ISPA Co.
Jet-Com Inc.
Kirk Industrial Sales Inc.
Machinists Inc.
Metallic Bonds Ltd.
Metallizing Equipment Co.

PVT Ltd.
Metco—Div. of Perkin-Elmer
National Coating Technologies Inc.
Plasma Coating Corp.
Plasma-Tec Inc.
Plasma Technology inc.
Powder Alloy Corp.
Praxair Surface Technologies Inc.
Purtech Coatings Co.
RDB Enterprises Inc.
Sermatech Technical Services
Southwest Aeroservice Inc.
Spray Masters Inc.
St. Gobain/Norton Industrial Ceramics Corp.
St. Louis Metallizing Co.
Stellite Coatings
Stork Cellramic Inc.
Sulzer Plasma Technik Inc.
Superior Arc Metalizing Co. Inc.
Surtech Inc.
Technetics Corp.
The Thermal Spray Laboratory
Thermal Coatings International Inc.
Thermal Spray Technologies Inc.
Union Carbide Coatings Service Corp.
Wall Colmonoy Corp.

Atmospheric Corrosion

A & A Co. Inc.
Adiabatics Inc.
Advanced Materials Technologies Inc.
Aerobraze Corp—Coatings
Air Products & Chemicals Inc.
Alamo Supply Co.
Alloy Carbide Co.
American Metallizing & Sandblasting Inc.
Amorphous Metals Technologies Inc. (AMTECH)
A.O. Smith Corp.—Protective Coatings Div.
Applied Polymer Systems Inc.
APS-Materials Inc.
ASB Industries Inc.
Bay State Abrasives—Div. Abrasive Industries Inc.
Bender Machine Services Ltd.
Brodeur Machine Co. Inc.
C & P Surface Technology
Cincinnati Thermal Spray Inc.
Cincinnati Thermal Spray—South
Douglas Call Co. Inc.
Ellison Surface Technologies Inc.
Engineered Coatings Inc.
Eutectic Corp.
Exline Inc.
Fisher Products Inc.
Flame Spray Engineering & Research
Flame Spray Coating Inc.
Flame-Spray Industries Inc.
Flame Spray Northwest
F.W. Gartner Thermal Spraying Co.
General Magnaplate Corp.
General Plasma Inc.
Heany Industries Inc.
HFW Industries Inc.
Hitemco
Hobart Tafa Technologies Inc.
Jet-Com Inc.
Kirk Industrial Sales Inc.
Metallic Bonds Ltd.
Metallizing Equipment Co. PVT Ltd.
Metco—Div. of Perkin-Elmer
National Coating Technologies Inc.
Plasma Coating Corp.
Plasma-Tec Inc.
Plasma Technology inc.
Powder Alloy Corp.
Purtech Coatings Co.
RDB Enterprises Inc.
Spray Masters Inc.
St. Louis Metallizing Co.
Stellite Coatings
Stork Cellramic Inc.
Sulzer Plasma Technik Inc.
Superior Arc Metalizing Co. Inc.
Surtech Inc.
Technetics Corp.
The Thermal Spray Laboratory
Thermal Coatings International Inc.
Thermal Spray Technologies Inc.
Wall Colmonoy Corp.

Abradable Applications

A & A Co. Inc.
Adiabatics Inc.
Advanced Materials Technologies Inc.
Aerobraze Corp—Coatings
Alamo Supply Co.
Alloy Carbide Co.
American Metallizing & Sandblasting Inc.
Amorphous Metals Technologies Inc. (AMTECH)
APS-Materials Inc.
ASB Industries Inc.
Bender Machine Services Inc.
Bender Machine Services Co. Ltd.
Brodeur Machine Co. Inc.
Cincinnati Thermal Spray Inc.
Cincinnati Thermal Spray—South
Ellison Surface Technologies Inc.
Engineered Coatings Inc.
Fisher Products Inc.
Flame Spray Engineering & Research
Flame Spray Coating Inc.

Flame-Spray Industries Inc.
F.W. Gartner Thermal Spraying Co.
General Magnaplate Corp.
General Plasma Inc.
Heany Industries Inc.
Hitemco
Hobart Tafa Technologies Inc.
The ISPA co.
Jet-Com Inc.
Kirk Industrial Sales Inc.
Metallizing Equipment Co. PVT Ltd.
Metco—Div. of Perkin-Elmer
National Coating Technologies Inc.
Plasma Coating Corp.
Plasma-Tec Inc.
Plasma Technology Inc.
Plasma Technology Inc.
Plastic Flamecoat Systems
Powder Alloy Corp.
Purtech Coatings
RDB Enterprises Inc.
Sermatech Technical Services
Southwest Aeroservise Inc.
Spray Masters Inc.
St. Gobain/Norton Industrial Ceramics Corp.
St. Louis Metallizing Co.
Sulzer Plasma Technik Inc.
Surtech Inc.
Technetics Corp.
The Thermal Spray Laboratory
Thermal Coatings International Inc.
Thermal Spray Technologies Inc.
Wall Colmonoy Corp.

Vacuum Plasma

Aerobraze Corp—Coatings
Alamo Supply Co.
APS-Materials Inc.
Electro-Plasma Inc.
Drexel University
Engineered Coatings Inc.
Hardface Alloys Inc.
Metallic Bonds Ltd.
Metco—Div. of Perkin-Elmer
Plasma-Technik AG
Powder Alloy Corp.
Sulzer Plasma Technik Inc.
Surtech Inc.
Tekna Plasma Systems Inc.
The Thermal Spray Laboratory

Appendix B

Available Powder Feedstocks

This compilation derived from supplier brochures is not a complete listing. Vendors identified under "Thermal Spray Process Supplies" in Appendix A should be contacted for further information.

Iron-Based Alloys

Ametek
Stainless steels
Iron aluminide intermetallic

Amorphous Metals Technologies
Armacor C†; forms amorphous surface layering response to abrasion

Anval
Type 420, 431, 316L, 316H, and 254SMO† stainless steels

F.J. Brodmann
Low- and high-carbon steels
316, 420, 413 stainless steels
Fe/Cr composite

CEAC
Type 304 and 316 stainless steels
Self-bonding composites based on low- and high-carbon and stainless steels

Dynamet
Fe-Al

Eutectic+Castolin
Self-bonding powder formulations

Hermann C. Starck

Hobart-Tafa
316 stainless (for HVOF)

Metco
Low-carbon steel
Type 316 and 431 stainless steels
High-carbon iron + Mo composite
Self-bonding low- and high-carbon steel composites
Self-bonding stainless composite
High-chromium stainless composite
Diamalloy 1008 blend with high-Cr and Mo (for HVOF)

Miller Thermal (Alloys International—"Plasmalloy")
304, 316, 410 stainless steels
Mild steel

Powder Alloy Corp.
Type 316 and 431 stainless steels
Low- and high-carbon steel composites with aluminum

Specialty Powders
FeCrAlYs

Stellite
Tristelle† stainless steels modified with cobalt and silicon
"High-grade" stainless steels for HVOF
Delcrome† iron-based alloys for low-temperature abrasion/erosion

Sulzer Surface Tech. ("Amdry")
Type 431 stainless
Fe-24Cr-8Al-0.5Y
Fe-37Ni-Al self-bonding clad product

Union Carbide
FeCrAlYs

Wear Management Services
Tribolite for metal-to-metal wear
Tribolite "One-Step" with bond-enhancing additive

Xform
Iron Aluminide (Fe_3Al)

Nickel-Based Alloys

Ametek
Ni-Cr-B
Ni-Cr-Mo
Nickel aluminide

Anval
- Ni-Al-(Mo)-(Cr)
- Alloys 600, 625, and 718
- Alloy† C-22

F.J. Brodmann
- Pure Ni
- Ni-Cr
- Ni-Cr-Mo-Al-Fe
- Ni-Al-(Mo)
- Ni-Cu
- Self-bonding composites
- Hardfacing formulations

CEAC
- Pure Ni
- Ni-Cr (some with self-fluxing additions)
- Prealloy and composite Ni-Al

Dynamet
- Ni-Cr
- Ni-Al
- M-CrAlY with M = Ni and/or Co
- Ni-Co-Cr-Al-Y-Hf

Eutectic+Castolin

Hermann C. Starck

Hobart-Tafa
- Pure Ni (for HVOF)
- Ni-Cr (for HVOF)
- Alloy 625 (for HVOF)
- Alloy 718 (for HVOF)
- Ni-Cr-Si-B (self-fluxing) (for HVOF)

Metco
- Pure Ni
- Ni-Cr
- Self-fluxing alloys
- Ni-Al composite
- Ni-Cr-Al
- Ni-Fe-Al-(Mo) composites
- Ni-Cr-Al-Co-yttria composite
- Type 700 for amorphous-like coatings

Miller Thermal (Alloys International "Plasmalloy")
- Pure Ni
- Ni-Cr
- Ni-Cr-Al composite
- Ni-Al-Cr-Fe-Mo
- Ni-Al-(Mo)
- Ni-Cr-Mo-W-(B)
- Ni-Fe-Co
- Ni-Cr-Fe-Si
- Hardfacing formulations

Powder Alloy Corp.
- Pure Ni
- Self-fluxing alloys
- Ni-Cr-(Fe)
- Ni-Al-(Cr)-(Mo); some are exothermic composites
- Superalloys
- NiCrAlYs

Specialty Powders
- Nickel-based superalloys
- Pure Ni
- Ni-Cr-(Al)
- Ni-Al-(Mo)
- NiCrAlYs

Stellite
- Deloro† Nistelle† powders, comparable to Haynes International Hastelloys†
- Tribaloy† T-700
- Ni-Cr for HVOF

Sulzer Surface Tech. ("Amdry")
- Pure Ni
- Ni-Al-(Mo), including clad and prealloy products
- Ni-Cr-(Al); Al may be alloyed or clad
- Ni-Cr-Al-(Y or Mo+Fe) gas atomized
- Ni-Co-Cr-Al-Ta-Y
- Proprietary NiCrAlYs
- Alloy 625
- Rene'†77 and 80

Union Carbide
- Ni-Cr
- Ni-Al
- NiCrAlYs
- Ni-based superalloys

Wall Colmonoy
- Ni-based hardfacing powders, mainly for spray-and-fuse and plasma transferred arc application

Xform
- nickel aluminides

Cobalt-Based Alloys

Anval
- Co-Cr-Ni-W-(B-Si)
- "Ultimet" alloy with high Mo

Ametek

F.J. Brodmann
- Co-Cr-Ni-W
- Hardfacing formulations

CEAC

Dynamet
- Co-Mo-Cr-Si
- Co-Cr-Al-Y or Hf
- Co-Cr-W-Ni-V
- Co-Cr-Ni-W

Hobart-Tafa
- Tribaloy† 800 (for HVOF)

Metco
- Co-Cr-W
- Co-Mo-Cr

Miller Thermal (Alloys International "Plasmalloy")
- Co-Cr-Ni-W
- Co-based hardfacing alloys

Powder Alloy Corp.

Specialty Powders
- Tribaloy[†] T-400, and T-800
- Coast Metal[†] 64
- Co-Cr-Ni-W
- CoCrAlYs
- Co-based superalloys

Stellite
- 13 Stellite[†] wear-resistant alloys, some appropriate for HVOF
- Tribaloy T-400 and T-800

Sulzer Surface Tech. ("Amdry")
- Tribaloy 400 and 800
- 17 other gas-atomized formulations

Union Carbide
- CoCrAlYs
- Co-based superalloys
- Alloys with high Mo, Cr, and W

Wall Colmonoy
- Wallex[†] C0-based hardfacing powders, mainly for spray-and-fuse and plasma transferred arc application

Refractory Metals

F.J. Brodmann
- Mo
- Ta

CEAC
- Pure Mo and blends with self-fluxing alloy
- "One-step" composite of Mo with Ni-Al

GTE
- Several Mo-based powders, including composites with Mo_2C and MoO_2, and blends with nickel alloy
- Tungsten

Metco
- Mo
- Self-fusing molybdenum blends
- W
- Ta

Miller Thermal (Alloys International "Plasmalloy")
- Mo
- Mo/NiCr(B)
- Ta
- Crystalline tungsten
- Titanium hydride

Powder Alloy Corp.
- Titanium and titanium alloys

Specialty Powders
- Mo

Stellite
- Mo for HVOF application

Sulzer Surface Tech. ("Amdry")
- Mo (four forms)
- Mo/molybdenum oxide
- Mo/NiCrBSiFe
- Nb
- Ti
- W

Union Carbide
- Mo

Other Metal and Semimetal Powders

Ametek
- Copper-nickel alloys
- Aluminum bronze

F.J. Brodmann
- Pure Sn, Zn, Al, Cu, Cr, Si
- Aluminum bronze
- Al-Si
- Cr-based composites with Al, Mo, Fe
- Molybdenum disilicide
- Titanium and titanium hydride for biomedical applications

CEAC
- Aluminum
- Al-Si
- Copper
- Cu-Ni-(In)
- Cu-P
- Aluminum bronze
- Aluminum nickel bronze

Dynamet
- Cu-Ni-(In)
- Aluminum bronze

Eutectic+Castolin
- Zinc- and aluminum-based for corrosion resistance
- Tin-based babbitt
- Bronze

Metco
- Al, Al-Si
- Zn
- Cu, Cu-Ni-(In)
- Aluminum bronze
- Babbitt

Micron Metals
- High-purity titanium and titanium alloys

Miller Thermal (Alloys International "Plasmalloy")
- Al, Al-Si
- Cu, Cu-Ni-(In)
- Aluminum bronze
- Cr-Si-B-Al-Ni

Powder Alloy Corp.
- Aluminum
- Al-Si
- Copper
- Cu-Ni-(In)
- Aluminum bronze

Specialty Powders
- Al-Si
- Cr
- Cu-Ni-(In)
- Aluminum bronze

Sulzer Surface Tech. ("Amdry")
- Al,Al-Si
- Cu, Cu-Ni-(In)
- Aluminum bronze

Union Carbide
- Al-Si
- Cu-Ni-(In)
- Aluminum bronze

Xform
- Titanium aluminides

Ceramics

Applied Powder Technologies
- High-purity ceramic powders by a novel spray drier/spray reactor process, including alumina, alumina-titania, chromia, chromia-silica, magnesium aluminate spinel and yttria, and calcia- or magnesia-stabilized zirconia

F.J. Brodmann
- Pure alumina, chromia, titania
- Alumina-titania
- Alumina-magnesia
- Alumina-zirconia
- Chromia-titania-(silica)
- Zirconia-magnesia
- Yttria- and calcia-stabilized zirconia
- Porcelain frit
- Superconductive oxide
- Semiconductive oxides
- Hydroxyl apatite for biomedical applications

CEAC
- Aluminum, chromium, titanium, and zirconium oxides
- Yttria-stabilized zirconia
- Alumina-titania
- Magnesium zirconate

Eutectic+Castolin
- Unspecified formulations suitable for Eutectic combustion torch application

GTE
- Zirconia-yttria (8% and 20%)

Hermann C. Starck

Hobart-Tafa

Metco
- Alumina, chromia, titania, zirconia
- Alumina-titania
- Chromia-silica
- Zirconia-titania-yttria
- Magnesium zirconate
- Magnesia-alumina spinel
- Yttria- and yttria/ceria-stabilized zirconia

Miller Thermal (Alloys International "Plasmalloy")
- Alumina, chromia, titania
- Alumina-titania
- Chromia-silica-titania
- Magnesium zirconate
- Calcia- and yttria-stabilized zirconia

Montreal Carbide
- 25 electofused ceramics, including yttria- and calcia-stabilized zirconia, alumina, alumina-titania, zirconia-alumina, chromia-titania, magnesium zirconate, spinel and other refractories, yttria, titania

Muscle Shoals Minerals
- Alumina
- Magnesium zirconate
- Alumina titanate
- Yttria- and calcia-stabilized zirconia

Norton
- Alumina (4 grades)
- Alumina-titania (9 grades up to 40% TiO_2)
- Magnesium aluminate (spinel)
- Magnesium zirconate
- Calcium zirconate
- Yttria- and calcia-stabilized zirconia
- Chromia (with and without silica and titania)

Powder Alloys Corp.
- Alumina, chromia, titania
- Alumina-titania (blends and composites)
- Chromia-silica
- Yttria- and calcia-stabilized zirconia
- Magnesium zirconate

Specialty Powders
Alumina, chromia, yttria
Alumina-titania
Zirconia-magnesia
Yttria- and ytterbia-stabilized zirconia

Sulzer Surface Tech. ("Amdry")
Alumina, chromia, titania
Alumina-titania
Alumina-zirconia
Chromia-silica-titania
Hafnia-yttria
Zirconia-magnesia
Yttria- and calcia-stabilized zirconia
Hydroxylapatite for biomedical applications

Union Carbide
Alumina, chromia
Alumina-titania
Zirconia-magnesia
Yttria-stabilized zirconia

Zircoa
Yttria-stabilized zirconia with novel porous morphology

Plastics

F.J. Brodmann
EAA
Nylon 11
Fluoropolymer ECTFE
Teflon† ETFA, FEP, PFA

Eutectic

Metco
polyester

Miller Thermal (Alloys International "Plasmalloy")
Epoxy
Nylon

National Coatings
EAA
MAA

Powder Alloys Corp.
aromatic polyester

UTP
EVOH
Nylon 11
EAA

Carbide- and Boride-Based Materials

Alloy Technology International
Ferrotic† titanium carbide with various matrix compositions including tool steel, martensitic stainless steels, and nickel-based alloys

Ametek

F.J. Brodmann
Pure chromium carbide and combined with NiCr
WC/Co
WC with Ni-based alloys
WC with chromium carbide and nickel

Eutectic+Castolin
Unspecified Diamax particles in nickel-based matrix

CEAC
Pure chromium carbide and blends with Ni-Cr
WC/Co
WC with self-fluxing Ni and Co alloy

GTE
Several cermets based on tungsten and chromium carbide, some made by novel plasma densification technique
Blends of WC/Co with Ni alloys
WC-TiC-Ni for erosive wear in alkaline solutions

Hermann C. Starck

Hobart-Tafa
WC/Co (for HVOF)
WC/Co blend with self-fluxing alloy (for HVOF)
Chromium carbide/Ni-Cr (for HVOF)

Metco
WC/Co
Pure chromium carbide, and combinations with NiCr and Ni
WC blends with self-fluxing alloys

Montreal Carbide
Boride-reinforced stainless steel matrix composites

Miller Thermal (Alloys International "Plasmalloy")
26 feedstocks based on WC
Pure chromium carbide
Chromium carbide/NiCr
Fused chromium carbide-tungsten carbide-NiCr-Mo

Powder Alloy Corp.
Chromium carbide, and blends with NiCr and bond enhancing powders
WC/Co (cast or sintered)
WC/Co+NiCrSi+NiAl blend

Specialty Powders

Pure chromium carbide, and combinations with Ni-Cr and FeCrAlY

WC with Co, Ni, or Co-Cr

WTiC-16%Ni

Stellite

Stelcar carbide containing alloys for plasma and HVOF

Chromium carbide/NiCr for HVOF

Sulzer Surface Tech. ("Amdry")

WC/Co

WC/Co-Cr

WC/Ni composite for HVOF spraying

Pure chromium carbide

Chromium carbide with NiCr (blend or aggolomerated and densified)

Chromium carbide blend with NiAl and self-fluxing Co-based alloy

Union Carbide

Pure chromium carbide, and combinations with Ni-Cr and FeCrAlY

WC with Co, Ni, or Co-Cr

WTiC-16%Ni

Wall Colmonoy

Ni-based formulations with tungsten carbide or chromium boride

Wallex No. 55 (50% WC in Co-Ni matrix)

Colmonoy No. 88 (bimetallic carbide in Ni-W-Cr matrix); available in HVOF, spray-and-fuse, and plasma transferred arc grades

Other Composites/ Blends/Compounds

F.J. Brodmann

Boron nitride with Al, Ni, alumina, or aluminum bronze

Teflon with Al, Al-Si, bronze, titanium boride, boron carbide, or boron nitride

Nickel/graphite

Aluminum/alumina

SiC with Al, Si, or alumina

CEAC

Magnesium zirconate composite with Ni-Cr,Ni-Al, or Ni-Cr-Al

Nickel-graphite

Metco

Boron nitride

Magnesium zirconate + NiCr

Nickel-graphite

Aluminum-graphite

Nickel-graphite

NiCrAl/Bentonite composite

Polyester blends with Al-Si and aluminum bronze

Bond coats for plastic substrates

Miller Thermal (Alloys International "Plasmalloy")

Nickel/graphite

Magnesium zirconate with Ni-Al, Ni-Cr, or Ni-Cr-Al

Powder Alloy Corp.

Polyester-Si

Nickel-clad graphite

Boron nitride + NiCrAl blend

Alumina + NiAl blends

Magnesia-zirconia + NiCr or NiAl blends

Sulzer Surface Tech. ("Amdry")

Al-Si composite with polyamide

Al-Si blend with polyester

Ni-graphite composites

Appendix C
Available Wire and Rod Feedstocks

Iron-Based Alloys

Amorphous Metals Technologies
"Armacor" alloys (form amorphous surface in response to abrasion)

F.J. Brodmann
18/5, 18/8, type 316 and 13% Cr stainless steels

ECS Metals
Carbon steel
Four grades of stainless

Metco
Comparable to Miller Thermal (see below)

Miller Thermal ("Mogul")
Carbon steel (4 levels of carbon)
Five grades of stainless steel
"Alcro" with 21-23%Cr, 3.5-5.5%Al
"Steel L" with Ni, Mo, Cr, Mn
"Tufton Chrome" with 1.5% Cr for abrasive wear (flame spray only)
"Tufton 500" with Cr and Mn for abrasion resistance

Hobart-Tafa
Low-, medium-, and high-carbon steels
18/5 and 18/8 stainless steels
13% Cr steel

Shaped Wire

Wall Colmonoy
Comparable to Miller Thermal (see above)

Nickel-Based Alloys

F.J. Brodmann
Pure nickel
Ni-Cr
Ni-Cu (Monel)

ECS Metals
Ni-Cr

Hobart-Tafa
Pure nickel
Ni-Cu (Monel)
Ni-Cr
Ni-Cr-Ti (TAFALOY-45CT for boiler protection)
Ni-Fe-Cr
Ni-Cr-Al (73MXC cored wire micromatrix composite)
Ni-Al (75B BondArc 5%Al prealloy)

Metco
Comparable to Miller Thermal (see below)

Miller Thermal ("Mogul")
Pure nickel
Ni-Cr
Ni-Cu (Monel)
Alloy 625
Ni-Al-Mo self-bonding material

Wall Colmonoy
Comparable to Miller Thermal (see above)

Copper-Based Alloys

F.J. Brodmann
Pure copper
Aluminum bronze
Phosphor bronze
Brass

ECS Metals
Pure copper
Aluminum bronze
Other bronzes
Brasses

Hobart-Tafa
Pure copper
Aluminum bronze
Phosphor bronze
Brass

Metco
Comparable to Miller Thermal (see below)

Miller Thermal ("Mogul")
Pure copper
Aluminum bronze
Phosphor bronze
Silicon bronze
Commercial bronze with 10% Zn
"Tobin" bronze (high zinc)
"Tufton" bronze (high zinc, plus tin and iron)
Yellow brass

Shaped Wire

Wall Colmonoy
Comparable to Miller Thermal (see above)

Aluminum-Based Materials

Alcan Aluminum ("Duralcan" Division)
Duralcan 90/10 aluminum/aluminum oxide composite

F.J. Brodmann
Pure aluminum

ECS Metals
Pure aluminum

Hobart-Tafa
Pure aluminum
Al-Si

Metco
Comparable to Miller Thermal (see below)

Miller Thermal ("Mogul")
Pure aluminum
Al-Si (4.5-6.0%Si)
Aluminum 1100 with Si+Fe = 1.0% max. for enhanced corrosion resistance than Al-Si

Shaped Wire

Wall Colmonoy
Comparable to Miller Thermal (see above)

Zinc-Based Alloys

F.J. Brodmann
Pure zinc

Hobart-Tafa
Pure zinc
Zinc-aluminum

Metco
Pure zinc

Miller Thermal
Pure zinc

Platt Bros.
Zinc
Zinc-aluminum

SNMI
"Dunois" Zn-Al

Wall Colmonoy
Pure zinc

Zinco
Pure zinc

Refractory Metals

F.J. Brodmann
Molybdenum
Titanium

ECS Metals
Molybdenum
Titanium alloys

GTE
Molybdenum

Metco
Molybdenum

Miller Thermal ("Mogul")
Molybdenum

Wall Colmonoy
Molybdenum

Carbide-Reinforced Materials

Amorphous Metals Technologies
Duocor† (W and Ti carbides in "hard alloy" matrix)

Ceramic Rods

F. J. Brodmann
Alumina
Chromia (purity 85% to 90%)

Miller Thermal ("Mogul")
Alumina
Zirconia-calcia
Chromia-silica-calcia-alumina-magnesia (83% chromia)

Norton ("Rokide")
Alumina
Chromia
Zirconia
Alumina-titania
Magnesium aluminate
Magnesium zirconate
Zirconium silicate
Yttria-stabilized zirconia

Other

F.J. Brodmann
Tin
Tin-zinc
Babbitt
Composites of copper and 302 stainless

Hobart-Tafa
Tin
Sn-Zn
Babbitt
Low-melting alloys for mold making

Metco
Tin
Babbitt

Miller Thermal ("Mogul")

Tin

Babbitt

SNMI

Novel "cord" feeds of powder in plastic cord; may include carbide-bearing materials

Wall Colmony

Tin

Babbitt

REFERENCES

1. Y. Arata, A. Ohmori, C.J. Li, "Study on the Structure of Plasma Sprayed Ceramic Coating Using Copper Electroplating," in *Proceedings of the International Symposium on Advanced Thermal Spraying Technology and Allied Coatings* (Osaka, Japan: High Temperature Society of Japan, 1988), pp. 205-210.

2. R.C. Tucker, *ASM International Thermal Spray Workshop*, held October 2, 1989, Materials Congress, Indianapolis, IN (Materials Park, OH: ASM International, 1989).

3. J.M. Houben, "Future Developments in Thermal Spraying," in *Thermal Spray Coatings—New Materials, Processes and Applications, Proceedings of the 2nd National Conference on Thermal Spray* (Materials Park, OH: ASM International, 1984), pp. 1-20.

4. B.A. Shaw, A.G. Morton, "Thermal Spray Coatings-Marine Performance and Mechanisms," in *Thermal Spray Technology—New Ideas and Processes, Proceedings of the National Thermal Spray Conference* (Materials Park, OH: ASM International, 1988), pp. 385-408.

5. R. McPherson, P. Cheang, "Microstructural Analysis of Ni-Al Plasma Sprayed Coatings," in *Proceedings of the 12th International Conference on Thermal Spraying*, paper no. 17 (London, UK: Welding Institute, 1989).

6. S.M. French, G.F. Vogel, "Improvement in Wear Resistance of Electrical Conductors Using Plasma Sprayed Coatings," in *Proceedings of the 12th International Conference on Thermal Spraying* (London, UK: Welding Institute, 1989), pp. 129-138.

7. R.O. Drossman, R.P. Krepski, "Tribolite Iron-Based Thermal Spray Coatings for the Control of Wear," in *Thermal Spray Coatings: Properties, Processes and Applications, Proceedings of the 4th National Thermal Spray Conference* (Materials Park, OH: ASM International, 1991), pp. 353-356.

8. R. Unger, "Comparison of Thermal Spray Bond Coats," in *Thermal Spray—Advances in Coatings Technology, Proceedings of the National Thermal Spray Conference* (Materials Park, OH: ASM International, 1987), pp. 365-370.

9. S. Sampath, H. Herman, S. Rangaswamy, "Ni-Al Re-Evaluated," in *Thermal Spray—Advances in Coatings Technology, Proceedings of the National Thermal Spray Conference* (Materials Park, OH: ASM International, 1987), pp. 47-55.

10. R.C. Tucker, "Plasma and Detonation Gun Deposition Techniques and Coating Properties," in *Deposition Technologies for Films and Coatings* (Park Ridge, NJ: Noyes Publications, 1982).

11. R.C. Tucker, Materials Congress, 1990, Detroit, MI (Materials Park, OH: ASM International, 1990).

12. A. Adamski, R. McPherson, "Cavitation Erosion Behavior of Several Thermally Sprayed Coatings," in *Advances in Thermal Spraying, Proceedings of the 11th International Thermal Spraying Conference* (Materials Park, OH: ASM International, 1986), pp. 523-532.

13. H. Steffens, U. Fisher, "Correlation Between Microstructure and Physical Properties of Plasma Sprayed Zirconia Coatings," in *Thermal Spray Technology—New Ideas and Processes, Proceedings of the National Thermal Spray Conference* (Materials Park, OH: ASM International, 1988), pp. 167-173.

14. J.E. Nerz, B.A. Kushner, A.J. Rotolico, "Microstructural Evaluation of Tungsten Carbide-Cobalt Coatings," in *Thermal Spray Coatings: Properties, Processes and Applications, Proceedings of the 4th National Thermal Spray Conference* (Materials Park, OH: ASM International, 1991), pp. 353-356.

15. G. Tremouilles, J.L. Derep, R. Portier, "Plasma Coated Metal-Zirconia Interface," in *Thermal Spray Coatings: Properties, Processes and Applications, Proceedings of the 4th National Thermal Spray Conference* (Materials Park, OH: ASM International, 1991), pp. 445-456.

16. V. Wilms, H. Herman, "Crystallography and Microstructure of Equilibrium and Metastable Phases Resulting from Plasma Spraying of Ceramic Coatings," in *Proceedings of the 8th International*

Thermal Spraying Conference (Miami Beach, FL: American Welding Society, 1976), pp. 236-242.

17. R.W. Smith, "Particle Melting Behavior in a D.C. Plasma Jet; Observations on Several Nickel Base Alloys," in *Thermal Spray Technology—New Ideas and Processes, Proceedings of the National Thermal Spray Conference* (Materials Park, OH: ASM International, 1988), pp. 123-128.

18. H. Herman, *Adv. Materials and Processes* 137, 4 (1990): p. 41.

19. M.L. Thorpe, *Adv. Materials and Processes* 134, 4 (1988): p. 69.

20. F. Hermanek, Materials Congress, held October 2, 1989, Indianapolis, IN (Materials Park, OH: ASM International, 1989).

21. E.J. Kubel, *Adv. Materials and Processes* 132, 6 (1987): p. 69

22. F. Longo, *Welding Journal* 45, 2(1966): p. 66.

23. E.J. Kubel, *Adv. Materials and Processes* 138, 6 (1990): pp. 24-32.

24. E.S. Hamel, *Materials Engineering* (Aug. 1986).

25. "Protective Coatings and Their Processing—Thermal Spray," Plasma Technik AG literature.

26. M.W. Poe, ASM Materials Congress, Detroit, MI (Materials Park, OH: ASM International, 1990).

27. R.W. Whitfield, 1990 National Thermal Spray Conference, Long Beach, CA (Materials Park, OH: ASM International, 1990).

28. H. Kreye et al., "Microstructure and Bond Strength of WC-Co Coatings Deposited by Hypersonic Flame Spraying (JET KOTE Process)," in *Advances in Thermal Spraying, Proceedings of the 11th International Thermal Spray Conference* (Materials Park, OH: ASM International, 1986), pp. 121-128.

29. H. Wang, P. Yu, J. Wen, "The Oxidation During Electric Arc Spraying and Its Control," in *Advances in Thermal Spraying, Proceedings of the 11th International Thermal Spray Conference* (Materials Park, OH: ASM International, 1986), pp. 771-776.

30. E. Erturk, H.D. Steffens, "Low Pressure Arc Spraying in Comparison with Low-Pressure Plasma Spraying," in *Advances in Thermal Spray, Proceedings of the 11th International Thermal Spraying Conference* (Materials Park, OH: ASM International, 1986), pp. 83-92.

31. S. Sampath, H. Herman, "Microstructure of Vacuum Plasma Sprayed Coatings," in *Thermal Spray Technology—New Ideas and Processes, Proceedings of the National Thermal Spray Conference* (Materials Park, OH: ASM International, 1988), pp. 1-9.

32. L.N. Moscowitz, "Application of HVOF Thermal Spraying to Solve Corrosion Problems in the Petroleum Industry," in *Thermal Spray: International Advances in Coatings Technology, Proceedings of the International Thermal Spray Conference* (Materials Park, OH: ASM International, 1992), pp. 611-618.

33. J.J. Kaiser, R.A. Miller, *Adv. Materials and Processes* 136 (Dec. 1989), p. 89.

34. Houk, Cheney, Cribbs, "Properties of Oxygen-Bearing Molybdenum Coatings," Proceedings of the 9th International Thermal Spray Conference (The Hague, Neth.: Nederlands Instituut voor Lastechniek, 1980), pp. 365-371.

35. D.A. Jager, D. Stover, W. Schlump, "High Pressure Plasma Spraying in Controlled Atmosphere Up to Two Bar," in *Thermal Spray: International Advances in Coatings Technology, Proceedings of the International Thermal Spray Conference* (Materials Park, OH: ASM International, 1992), pp. 69-80.

36. R.C. Tucker, "Plasma and Detonation Gun Deposition Techniques and Coating Properties," in *Deposition Technologies for Films and Coatings* (Park Ridge, NJ: Noyes Publications, 1982).

37. A. Kay, "Jet Kote Applications for Corrosion Protection," in *Thermal Spray—Advances in Coatings Technology, Proceedings of the National Thermal Spray Conference* (Materials Park, OH: ASM International, 1987), pp. 393-399.

38. *Thermal Spraying Practice, Theory, and Applications* (Miami, FL: American Welding Society, 1985).

39. M. Thorpe, *Chem. Tech.* (June 1982): p. 364.

40. Drzeniek, H.D. Steffens, "Cored Tube Wires for Arc and Flame Spraying," in *Thermal Spray—Advances in Coatings Technology, Proceedings of the National Thermal Spray Conference* (Materials Park, OH: ASM International, 1987), pp. 33-40.

41. T.F. Bernecki, "Coatings for Aggressive Environments," workshop by Northwestern University Basic Industries Research Laboratories, presented at the ASM Materials Congress, Indianapolis, IN, October, 1989 (Materials Park, OH: ASM International, 1989).

42. Eagle Arc 600 product literature.

43. D.R. Marantz, "Electromagnetically Coalesced Multi-Arc Plasma Torch with True Axial Powder Feed," in *Thermal Spray Research and Applications, Proceedings of the 3rd National Thermal Spray Conference* (Materials Park, OH: ASM International, 1990), pp. 443.

44. ASM *Metals Handbook*, "Corrosion," 9th ed., vol. 13 (Materials Park, OH: ASM International, 1987).

45. D. Gansert, G. Bancki, H. Herman, "Fare-Gun Formed Wear Resistant Coatings," in *Thermal Spray—Advances in Coatings Technology, Proceedings of the National Thermal Spray Conference* (Materials Park, OH: ASM International, 1987), pp. 203-209.

46. K.V. Rao, D.A. Somerville, D.A. Lee, "Properties and Characterization of Coatings Made Using Jet Kote Thermal Spray Technique," in *Advances in Thermal Spraying, Proceedings of the 11th International Thermal Spraying Conference* (Montreal, Canada: Welding Institute of Canada, 1986), pp. 873-882.

47. M.R. Dorfman et al., "A Technical Assessment of HVOF," 12th International Conference on Thermal Spraying, paper no. 17 (London, UK: The Welding Institute, 1989).

48. D.J. Gansert, H. Herman, "Jet-Kote Sprayed Machine Element Coatings," in *Thermal Spray Technology—New Ideas and Processes, Proceedings of the National Thermal Spray Conference* (Materials Park, OH: ASM International, 1988), pp. 139-146.

49. H. Kreye et al., "Microstructure and Bond Strength of WC-Co Coatings Deposited by Hypersonic Flame Spraying," in *Advances in Thermal Spraying, Proceedings of the 11th International Thermal Spray Conference* (Montreal, Canada: Welding Institute of Canada, 1986), pp. 121-128.

50. D.W. Parker, G.L. Kutner, *Adv. Materials and Processes* 139, 4 (1991): p. 68.

51. H. Herman, *Adv. Materials and Processes* 137, 1 (1990): p. 59.

52. H.D. Steffens, Hohle, "Further Development of the Thermal Spraying Technique by Low Pressure Plasma Spraying with Transferred Arc," 9th International Thermal Spray Conference (The Hague, Neth.: Nederlands Instituut voor Lastechniek, 1980), pp. 420-429.

53. H.D. Steffens, M. Wewel, "Recent Developments in Vacuum Arc Spraying," in *Thermal Spray Technology—New Ideas and Processes, Proceedings of the National Thermal Spray Conference* (Materials Park, OH: ASM International, 1988), pp. 33-38.

54. E. Lugscheider, I. Rass, *J. Thermal Spray Technology* 1, 1 (March 1992), pp. 49-55.

55. D.A. Jager, D. Stover, W. Schlump, "High Pressure Plasma Spraying in Controlled Atmosphere Up to Two Bar," in *Thermal Spray: International Advances in Coatings Technology, Proceedings of the International Thermal Spray Conference* (Materials Park, OH: ASM International, 1992), pp. 69-80.

56. P.G. Tsantrizos, "The Reactive Spray Forming Production of Titanium Aluminides in the Tail Flame of a D.C. Plasma Torch," in *Thermal Spray: International Advances in Coatings Technology, Proceedings of the International Thermal Spray Conference* (Materials Park, OH: ASM International, 1992), pp. 195-199.

57. W. Elger et al., "Thermal Sprayed Plasma Wire Applications for the U.S. Navy," 12th International Conference on Thermal Spraying, paper no. 67 (London, UK: Welding Institute, 1989).

58. M.L. Thorpe, W.R. Kratochvil, "Inductive Plasma Heating of Powder and Gases," 12th International Conference on Thermal Spraying, paper no. 102 (London, UK: Welding Institute, 1989).

59. M.L. Thorpe, W.R. Kratochvil, "Inductive Heating for a New Type of Plasma," in *Thermal Spray Technology—New Ideas and Processes, Proceedings of the National Thermal Spray Conference* (Materials Park, OH: ASM International, 1988), pp. 9-18.

60. X.L. Jiang et al., "Induction Plasma Spraying of Refractory Materials," in *Thermal Spray: International Advances in Coatings Technology, Proceedings of the International Thermal Spray Conference* (Materials Park, OH: ASM International, 1992), pp. 39-44.

61. P. Tajchman, L. Bula, "Properties of Coatings Applied by Plasma with Water Stabilization on Large Size Parts," 12th International Conference on Thermal Spraying, paper no. 68 (London, UK: Welding Institute, 1989).

62. J.H. Malinka, D.J. Varacalle, W.L. Riggs, "Thermal Sprayed Titania Coatings Using Liquid Stabilized Plasma Technology," in *Thermal Spray: International Advances in Coatings Technology, Proceedings of the International Thermal Spray Conference* (Materials Park, OH: ASM International, 1992), pp. 87-92.

63. P.A. Kammer, 1990 National Thermal Spray Conference, Long Beach, CA (Materials Park, OH: ASM International, 1990).

64. D.R. Marantz, "State of the Art Arc Spray Technology," in *Thermal Spray Research and Applications, Proceedings of the 3rd National Thermal Spray Conference* (Materials Park, OH: ASM International, 1990), pp. 113-118.

65. K. Altorfer, D. Marantz, U.S. Patent No. 4,269,867 (May 26, 1981).

66. K.H. Busse, "Arc Spraying of Cored Wires," 12th International Conference on Thermal Spraying, paper no. 36 (London, UK: Welding Institute, 1989).

67. S. Dallaire, H. Levert, "Phenomenology of Reactive Core Wire Arc Spraying," in *Thermal Spray: International Advances in Coatings Technology, Proceedings of the International Thermal Spray Conference* (Materials Park, OH: ASM International, 1992), pp. 207-212.

68. L. Moscowitz, Amoco, correspondence to author, 1990.

69. L. Moskowitz, D. Lindley, U.S. Patent No. 4,869,936 (Sept. 26, 1989).

70. J.B. Browning, "And Now, Hypersonic Velocity Impact Fusion Spraying," in *Thermal Spray: International Advances in Coatings Technology, Proceedings of the International Thermal Spray Conference* (Materials Park, OH: ASM International, 1992), pp. 123-125.

71. A.R. Nicoll, H. Eschnauer, "Quality Assurance of Plasma Spray Powders," in *Advances in Thermal Spraying: Proceedings of the 11th International Thermal Spray Conference* (Montreal, Canada: Welding Institute of Canada, 1986), pp. 289-298.

72. S.G. Lee, S. Safai, "Automation of the Thermal Spray Process," in *Advances in Thermal Spraying, Proceedings of the 11th International Thermal Spray Conference* (Montreal, Canada: Welding Institute of Canada, 1986), pp. 197-208.

73. W. Mayr, "Rapid Optimization of Spraying Parameters by Means of an Automated Laser Doppler Measuring Equipment," in *Advances in Thermal Spraying, Proceedings of the 11th International Thermal Spray Conference* (Montreal, Canada: Welding Institute of Canada, 1986), pp. 221-230.

74. *Adv. Materials and Processes* 140 (August 1991): p. 9.

75. *Adv. Materials and Processes* 140 (July 1991): pp. 30-32.

76. M.R. Dorfman, J.D. Reardon, "Spherical Ceramic Powders for Thermal Spraying," in *Advances in Thermal Spraying, Proceedings of the 11th International Thermal Spray Conference* (Montreal, Canada: Welding Institute of Canada, 1986), pp. 241-250.

77. C.C. Berndt, J.H. Yi, "Composite Thermal Sprayed Coatings," 12th International Conference on Thermal Spraying, paper no. 2 (London, UK: Welding Institute, 1989).

78. S.L. Van Doren, M.A. Riley, M.E. Marousek, "Boride-Whisker-Reinforced Coatings," in *Thermal Spray Technology—New Ideas and Processes, Proceedings of the National Thermal Spray Conference* (Materials Park, OH: ASM International, 1988), pp. 465-469.

79. X. Guo, H. Herman, S. Rangaswamy, "Cavitation-Erosion of Plasma-Sprayed WC/Co," in *Advances in Thermal Spraying, Proceedings of the 11th International Thermal Spray Conference* (Montreal, Canada: Welding Institute of Canada, 1986), pp. 37-46.

80. M.E. Vinayo, F. Kassabji, P. Fauchais, "Structural and Mechanical Properties of Co-WC, Ni-WC Deposits," in *Advances in Thermal Spraying, Proceedings of the 11th International Thermal Spray Conference* (Montreal, Canada: Welding Institute of Canada, 1986), pp. 93-100.

81. S. Rangaswamy, H. Herman, "Metallurgical Characterization of Plasma Sprayed WC-Co Coatings," in *Advances in Thermal Spraying, Proceedings of the 11th International Thermal Spray Conference* (Montreal, Canada: Welding Institute of Canada, 1986), pp. 101-110.

82. E. Lugscheider et al., "Influence of Different Plasma Spray Processes and Various Types of Stabilized Zirconia on the Morphology of Thermal Barrier Coatings," in *Advances in Thermal Spraying, Proceedings of the 11th International Thermal Spray Conference* (Montreal, Canada: Welding Institute of Canada, 1986), pp. 261-268.

83. D.B. Ott, B.A. Kushner, "Spherical Reacted YSZ Powders for Improved Thermal Spray Properties," in *Thermal Spray Research and Applications, Proceedings of the 3rd National Thermal Spray Conference* (Materials Park, OH: ASM International, 1990), pp. 93-98.

84. W. Milewski, M. Sartowski, "Some Properties of Coatings Arc-Sprayed in Nitrogen or Argon Atmosphere," in *Advances in Thermal Spraying, Proceedings of the 11th International Thermal Spraying Conference* (Montreal, Canada: Welding Institute of Canada, 1986), pp. 467-474.

85. R.O. Drossman, R.P. Krepski, "Tribolite Iron-Based Thermal Spray Coatings for the Control of Wear," in *Thermal Spray Coatings: Properties, Processes and Applications, Proceedings of the 4th National Thermal Spray Conference* (Materials Park, OH: ASM International, 1991), pp. 353-356.

86. D.M. Scruggs, "Coatings for Wear and Corrosion Utilizing Metamorphic Alloys," in *Thermal Spray Coatings: Properties, Processes and Applications, Proceedings of the 4th National Thermal Spray Conference* (Materials Park, OH: ASM International, 1991), pp. 147-152.

87. H.D. Steffens, H. Drzeniek, Z. Babiak, "Influence of Rare Earth Elements on Properties of Electric-Arc Sprayed Coatings Using Cored Wires," in *Thermal Spray Technology—New Ideas and Processes, Proceedings of the National Thermal Spray Conference* (Materials Park, OH: ASM International, 1988), pp. 325-330.

88. H.D. Steffens, H. Drzeniek, Z. Babiak, "Wear Resistant Composite Coatings Produced by Arc Spraying Using Cored Wires," in *Thermal Spray Technology—New Ideas and Processes, Proceedings of the National Thermal Spray Conference* (Materials Park, OH: ASM International, 1988), pp. 331-336.

89. B.A. Shaw, A.M. Leimkuhler, P.J. Moran, "Corrosion Performance of Aluminum and Zinc-Aluminum Thermal Spray Coatings in Marine Environments," in *Testing of Metallic and Inorganic Coatings*, ASTM STP 947 (Philadelphia, PA: ASTM, 1987), pp. 246-264.

90. B. Arsenault et al., "Zinc-Nickel Coatings for Improved Adherence and Corrosion Resistance," in *Thermal Spray Technology—New Ideas and Processes, Proceedings of the National Thermal Spray Conference* (Materials Park, OH: ASM International, 1988), pp. 409-416.

91. R.P. Krepski, N.G. Feige, "Zinc-Nickel Thermal Spray Coating for Mitigation of Stress Corrosion Cracking and Corrosion of Steel in Concrete," 1992 International Thermal Spray Conference, (Materials Park, OH: ASM International, 1992).

92. R.P. Krepski, "Zinc-Silicate Thermal Spray Coating for Corrosion Control of Steel in Concrete," in *Thermal Spray Coatings: Properties, Processes and Applications, Proceedings of the 4th National Thermal Spray Conference* (Materials Park, OH: ASM International, 1991), pp. 433-438.

93. R.C. Cobb, S.J. Harris, T. Lester, "The Development of Ni-Cr-B-Si Coatings by Electric Arc Spraying of Cored Wires," 12th

International Conference on Thermal Spraying, paper no. 105 (London, UK: Welding Institute, 1989).

94. R.G. Tallent, A.L. Plumley, *TAPPI J.* 52, 10 (1969): p. 1955.

95. C. Lane, R. Sulit, "Extended Service Life with an Aluminum Composite Non-Skid Coating," in *Thermal Spray Research and Applications, Proceedings of the 3rd National Thermal Spray Conference* (Materials Park, OH: ASM International, 1990), pp. 105-108.

96. E. Lugscheider, A. Nisch, T. Weber, "Thermal Spraying of Boron Carbide Containing Compounds," 12th International Conference on Thermal Spraying, paper no. 41 (London, UK: The Welding Institute, 1989).

97. R.L. McCaw, R.A. Hays, "Application of Commercially Pure Titanium Coatings Using the High Velocity Oxy-Fuel Thermal Spray Process Under Atmospheric Conditions," in *Thermal Spray: International Advances in Coatings Technology, Proceedings of the International Thermal Spray Conference* (Materials Park, OH: ASM International, 1992), pp. 881-886.

98. H.D. Steffens, M. Wewel, "Vacuum Arc Spraying—An Advanced Coating Technology," 12th International Conference on Thermal Spraying, paper no. 61 (London, UK: The Welding Institute, 1989).

99. C. d'Angelo, H. El Joundi, *Adv. Materials and Processes* 134, 6 (Dec. 1988): p. 41.

100. C. Berndt, correspondence to author, 1992.

101. M.R. Dorfman, J.D. Reardon, "Spherical Ceramic Powders for Thermal Spraying," in *Advances in Thermal Spraying, Proceedings of the 11th International Thermal Spray Conference* (Montreal, Canada: Welding Institute of Canada, 1986), pp. 241-250.

102. G.N. Heintze, R. McPherson, "Plasma-Sprayed Transformation Toughened ZrO_2 Coatings," 12th International Conference on Thermal Spraying, paper no. 5 (London, UK: The Welding Institute, 1989).

103. H.G. Wang, H. Herman, "Plasma-Sprayed Cordierite: Dielectric and Electrical Properties," in *Thermal Spray Technology—New Ideas and Processes, Proceedings of the National Thermal Spray Conference* (Materials Park, OH: ASM International, 1988), pp. 345-352.

104. L. McQueen, *J. Protective Coatings and Linings* 5, 2 (1988): p. 3.

105. T.W. Glass, J.A. DePay, "Protective Thermoplastic Powder Coating Specifically Designed Adhesive Polymers," in *Thermal Spray Coatings: Properties, Processes and Applications, Proceedings of the 4th National Thermal Spray Conference* (Materials Park, OH: ASM International, 1991), pp. 345-352.

106. C. Reynolds, National Coatings, correspondence to author, 1992.

107. F. Longo, correspondence to author, 1992.

108. R. Unger, "Arc Sprayed Nickel-Aluminum," in *Thermal Spray Technology—New Ideas and Processes, Proceedings of the National Thermal Spray Conference* (Materials Park, OH: ASM International, 1988), pp. 321-324.

109. S. Kozerski, "The Oxide Reduction in Plasma Sprayed Ni-Al Layers," in *Advances in Thermal Spraying, Proceedings of the 11th International Thermal Spray Conference* (Montreal, Canada: Welding Institute of Canada, 1986), pp. 845-853.

110. J. Wen, Z. Zhang, "An Investigation of the Self-bonding Mechanism of Aluminum Bronze Coatings Sprayed by the Electric Arc Process," in *Advances in Thermal Spraying, Proceedings of the 11th International Thermal Spray Conference* (Montreal, Canada: Welding Institute of Canada, 1986), pp. 169-176.

111. AWS C2.2-67, "Recommended Practices for Metallizing with Aluminum and Zinc for Protection of Iron and Steel" (New York, NY: American Welding Society, 1967).

112. AWS C2.14-74, "Corrosion Tests of Flame-Sprayed Coated Steel; 19-Year Report" (Miami, FL: American Welding Society, 1974).

113. *Adv. Materials and Processes* 140 (Aug. 1991): p. 12.

114. T. Race, V. Hock, A. Beitelman, *J. Protective Coatings and Linings* (Aug. 1989): p. 37.

115. R. Unger, Hobart-TAFA, correspondence to author, 1992.

116. W.I. Pollock, B.J. Moniz, eds., *Process Industries Corrosion—The Theory and Practice* (Houston, TX: NACE, 1986).

117. L. Tokoi, "Solutions for Wear and Corrosion," Surface Alloying and Wearfacing Services Company Report, September 1989.

118. W. Milewski et al., "Influence of Tightening Means on Structure and Some Properties of Plasma Sprayed Coatings," in *Advances in Thermal Spraying, Proceedings of the 11th International Thermal Spray Conference* (Montreal, Canada: Welding Institute of Canada, 1986), pp. 577-584.

119. B.G. McMordie, "Corrosion and Erosion Resistant Composite Coating for Turbomachinery Blading," in *Thermal Spray Technology—New Ideas and Processes, Proceedings of the National Thermal Spray Conference* (Materials Park, OH: ASM International, 1988), pp. 417-422.

120. A. Adamski, R. McPherson, "Laser Processing of Thermally Sprayed Coatings," in *Advances in Thermal Spraying, Proceedings of the 11th International Thermal Spray Conference* (Montreal, Canada: Welding Institute of Canada, 1986), pp. 555-562.

121. M. Havrda et al., "Laser Treatment of Plasma Sprayed Zirconium Silicate Coatings," in *Advances in Thermal Spraying, Proceedings of the 11th International Thermal Spray Conference* (Montreal, Canada: Welding Institute of Canada, 1986), pp. 569-576.

122. H. Ito et al., "Post-Treatment of Plasma Sprayed WC-Co Coatings by Hot Isostatic Pressing," in *Thermal Spray Research and Applications, Proceedings of the 3rd National Thermal Spray Conference* (Materials Park, OH: ASM International, 1990), pp. 233-239.

123. W. Milewski, Sicinaki, "The Effects of Metallic Dopes on Corrosion Resistance of Plasma Sprayed Coatings Made of Metal Oxides," 9th International Thermal Spray Conference (The Hague, Neth.: Nederlands Instituut voor Lastechniek, 1980), pp. 382-384.

124. K. I'Anson et al., "Use of the Shot Peening Process for Surface Modification of Thermal Sprayed Coatings," in *Thermal Spray Coatings: Properties, Processes and Applications, Proceedings of the 4th National Thermal Spray Conference* (Materials Park, OH: ASM International, 1991), pp. 139-146.

125. V. Meringolo, ed., *Thermal Spray Coatings* (Atlanta, GA: TAPPI Press, 1983).

126. G.R. Bell, "The Finish Machining of Sprayed Metal Coatings," in *Advances in Thermal Spraying, Proceedings of the 11th International Thermal Spraying Conference* (Montreal, Canada: Welding Institute of Canada, 1986), pp. 545-554.

127. L. Cronjager, P. Muller, U. Fischer, "Investigations About the Machinability of Arc-Sprayed Steel Coatings," in *Advances in Thermal Spraying, Proceedings of the 11th International Thermal Spray Conference* (Montreal, Canada: Welding Institute of Canada, 1986), pp. 863-872.

128. R. Massad, "Diamond Wheel Grinding of Thermal Spray Materials," in *Thermal Spray Coatings—New Materials, Processes and Applications, Proceedings of the 2nd National Conference on Thermal Spray* (Materials Park, OH: ASM International, 1984), pp. 139-146.

129. H. Herman, ASM Materials Congress, Detroit (Materials Park, OH: ASM International, 1990).

130. K. Ishikawa et al., "Protection from High Temperature Corrosion with a Multi-Layer Thermal Spray Coating," in *Thermal Spray: International Advances in Coatings Technology, Proceedings of the International Thermal Spray Conference* (Materials Park, OH: ASM International, 1992), pp. 859-864.

131. H.D. Steffens, Hohle, "Dangerous Events Caused by Thermal Spraying," 9th International Thermal Spray Conference (The Hague, Neth.: Nederlands Instituut voor Lastechniek, 1980), pp. 432-437.

132. Panel Discussion on Thermal Spray Automation, ASM Materials Congress, Indianapolis, IN (Materials Park, OH: ASM International, 1989).

133. D.B. Fowler, W. Riggs, J.C. Russ, *Adv. Materials and Processes* 138, 5 (Nov.1990): pp. 41-52.

134. "Characterizing Thermal-Spray Coatings," *Adv. Materials and Processes* 141, 5 (May 1992): pp. 23-27.

135. W.H. Thomason, R.G. Ivie, "Experience with a Hydraulic Adhesion Tester for Quality Control of Thermal-Sprayed Aluminum Coatings for Long Term Submerged Service in the Ocean," in *Thermal Spray Research and Applications, Proceedings of the 3rd National Thermal Spray Conference* (Materials Park, OH: ASM International, 1990), pp. 285-290.

136. S.D. Brown, B.A. Chapman, G.P. Wirtz, "Fracture Kinetics and the Mechanical Measurement of Adherence," in *Thermal Spray Technology—New Ideas and Processes, Proceedings of the National Thermal Spray Conference* (Materials Park, OH: ASM International, 1988), pp. 147-158.

137. C. Berndt, *J. Mater. Eng.* 12, 2 (1990): pp. 151-158.

138. C. Berndt, "Material Property Measurements on Thermal Barrier Coatings," 1988 Gas Turbine and Aeroengine Congress, paper no. 88-GT-277 (Amsterdam, Neth.: ASME, 1988).

139. C. Berndt et al., "Material Property Variations in Thermally Sprayed Coatings," in *Thermal Spray Coatings: Properties, Processes and Applications, Proceedings of the 4th National Thermal Spray Conference* (Materials Park, OH: ASM International, 1991), pp. 199-204.

140. E.H. Meyer, K.J. Pohl, "Further Development of the Holographic Sound Field Imaging Method for Testing Thermal Sprayed Coatings," 12th International Conference on Thermal Spraying, paper no. 33 (London, UK: Welding Institute, 1989).

141. H. Reiter, D. Almond, M.P. Patel, "The Non-Destructive Evaluation of Surface Coating by Photothermal Imaging," in *Thermal Spray Coatings—New Materials, Processes and Applications, Proceedings of the 2nd National Conference on Thermal Spray* (Materials Park, OH: ASM International, 1984), pp. 119-124.

142. D.R. Green et al., "System for NDE of Thermal Spray Coating Bonds," in *Thermal Spray Coatings—New Materials, Processes and Applications, Proceedings of the 2nd National Conference on Thermal Spray* (Materials Park, OH: ASM International, 1984), pp. 125-134.

143. J.D. Morris et al., "Developments in Thermal Wave Non-Destructive Testing Systems for Thermal Spray Coatings," 12th International Conference on Thermal Spraying, paper no. 99 (London, UK: Welding Institute, 1989).

144. C.F. Lewis, *Mater. Eng.* (Aug. 1988): pp. 45-48.

145. W.A. Nelson, National Thermal Spray Conference, Long Beach, CA (Materials Park, OH: ASM International, 1990).

146. R. Travis, G. Ginther, C. Zanis, "Infrared Thermal Wave Non-Destructive Evaluation of Thermal Sprayed Coatings," in *Advances in Thermal Spraying, Proceedings of the 11th International Thermal Spray Conference* (Montreal, Canada: Welding Institute of Canada, 1986), pp. 309-314.

147. P. Siitonen et al., "Electrochemical Method for Evaluating the Corrosion Resistance and Porosity of Thermal Sprayed Coatings," in *Thermal Spray: International Advances in Coatings Technology, Proceedings of the International Thermal Spray Conference* (Materials Park, OH: ASM International, 1992), pp. 853-858.

148. D.H. Harris, "Surface Architecture in Processing Industries," in *Advances in Materials Technology for Process Industries' Needs*, Proceedings of the NACE Conference, Atlanta, GA, 1984 (Houston, TX: NACE International, 1985).

149. C. Nusum, "Applications in Thermal Spraying," in *Thermal Spray Technology—New Ideas and Processes, Proceedings of the National Thermal Spray Conference* (Materials Park, OH: ASM International, 1988), pp. 449-454.

150. Smolak, Kammer, poster session in *Thermal Spray: International Advances in Coatings Technology, Proceedings of the International Thermal Spray Conference* (Materials Park, OH: ASM International, 1992).

151. A. Wensley, MacMillan Bloedel Ltd., correspondence to author, May, 1992.

152. J. Duke, A. Gaines, *Chem. Processing* (Oct. 1983): p. 82.

153. D. Singbeil, *Pulp and Paper, Canada* 87, 3 (1986): p. 93.

154. C.E. Guzi, *Pulp and Paper Industry Corrosion Problems*, vol. 4 (Houston, TX: NACE, 1983), p. 37.

155. C.E. Guzi, Proceedings of the 1985 TAPPI Engineering Conference (Atlanta, GA: TAPPI Press, 1983), p. 27.

156. D.R. McIntyre, C.P. Dillon, *Guidelines for Preventing SCC in the CPI*, MTI Publication #15 (Houston, TX: NACE, 1985).

157. H. Van Droffelaar, G. B. Hudson, "Stress Chloride Cracking, A Past Threat?," *Canadian Chemical Processing* (February 1966): pp. 78-80; *Canadian Chemical Processing* (March 1966): pp. 63-66.

158. N. Tani et al., "Protective Spray Coating Against Stress Corrosion Cracking of Austenitic Stainless Steel," in *Thermal Spray Research and Applications, Proceedings of the 3rd National Thermal Spray Conference* (Materials Park, OH: ASM International, 1990), pp. 211-214.

159. L. McGuire, NACE Regional Corrosion Conference, held Sept. 23, 1990, Detroit, MI (Houston, TX: NACE, 1990).

160. K.G. Budinski, *Surface Engineering for Wear Resistance* (Englewood Cliffs, NJ: Prentice-Hall, 1988).

161. P. Crook, *Materials Performance* 31, 2 (1991): p. 64.

162. J. Ritchie, "Selecting Thermal Spray Coatings for the Repair of Equipment Parts by the Job Shop," in *Thermal Spray Coatings—New Materials, Processes and Applications, Proceedings of the 2nd National Conference on Thermal Spray* (Materials Park, OH: ASM International, 1984), pp. 87-90.

163. A. Kay, "Successful Hypersonic Spray Applications in the Steel Industry," in *Thermal Spray Technology—New Ideas and*

Processes, Proceedings of the National Thermal Spray Conference (Materials Park, OH: ASM International, 1988), pp. 443-448.

164. H. Kreye, "Optimization and Control of the Spray Conditions in the Jet Kote Process," in *Thermal Spray Technology—New Ideas and Processes, Proceedings of the National Thermal Spray Conference* (Materials Park, OH: ASM International, 1988), pp. 39-46.

165. R. Bamola, ASM Materials Congress, Detroit, MI (Materials Park, OH: ASM International, 1990).

166. H. Kreye, "Characteristics of Coatings Produced by High Velocity Flame Spraying," 12th International Conference on Thermal Spraying, paper no. 24 (London, UK: Welding Institute, 1989).

167. "Jet Kote II Surfacing Systems Applications Notebook," Stellite product literature, 1990.

168. R.C. Tucker, "Coatings for Centrifugal Pump Components," 3rd International Pump Symposium, College Station, TX (1989).

169. R.B. Dawson, C.B. Thompson, *Materials Performance* 30, 11 (1991): p. 75.

170. D.M. Scruggs, "Coatings for Wear and Corrosion Utilizing Metamorphic Alloys," in *Thermal Spray Coatings: Properties, Processes and Applications, Proceedings of the 4th National Thermal Spray Conference* (Materials Park, OH: ASM International, 1991), pp. 147-152.

171. E.S. Hamel, "Ceramic Coatings: More than Just Wear Resistant," *Mater. Eng.* (Aug. 1986).

172. R.B. Heimann, D. Lamy, T.N. Sopkow, "Parameter Optimization of Alumina-Titania Coatings by a Statistical Experimental Design," in *Thermal Spray Research and Applications, Proceedings of the 3rd National Thermal Spray Conference* (Materials Park, OH: ASM International, 1990), pp. 491-496.

173. R.W. Smith et al., "The Structure and Properties of Plasma Sprayed TiC Dispersion Hardened Coatings," in *Thermal Spray Technology—New Ideas and Processes, Proceedings of the National Thermal Spray Conference* (Materials Park, OH: ASM International, 1988), pp. 299-306.

174. R.W. Smith et al., "A New Iron-Based Thermal Spray Coating for Wear Resistance," in *Thermal Spray: International Advances in Coatings Technology, Proceedings of the International Thermal Spray Conference* (Orlando, FL: ASM International, 1992), pp. 653-660.

175. K.S. Fessenden, Z. Zurecki, T.P. Slavin, "Sliding Wear Resistance of Electric Arc Sprayed Coatings," *Thermal Spray Coatings: Properties, Processes and Application, Proceedings of the 4th National Thermal Spray Conference* (Materials Park, OH: ASM International, 1991), pp. 81-88.

176. G.W. Wilkes, Amoco, correspondence to author, 1990.

177. G.W. Wilkes, ASM Materials Congress, Detroit MI (Materials Park, OH: ASM International, 1990).

178. R. Travis et al., "Thermal Spray Coatings for Shipboard Steam Valve Stems," in *Thermal Spray: International Advances in Coatings Technology, Proceedings of the International Thermal Spray Conference* (Materials Park, OH: ASM International, 1992), pp. 601-603.

179. L. Tokoi, "High Technology Solutions for Wear and Corrosion," Surface Alloying and Wearfacing Services (Houston, TX) company report, 1990.

180. Cummings, Brown, "Some Factors Influencing the Use of Surface Coatings in Sea Water," 9th International Thermal Spray Conference (The Hague, Neth.: Nederlands Instituut voor Lastechniek, 1980), pp. 400-409.

181. P.H. Thorpe, "Experience with Fabrication and Maintenance of Stainless Steel Clad Pulp Digesters," in *Pulp and Paper Industry Corrosion Problems, Proceedings of the 4th International Symposium* (Stockholm, Sweden: Swedish Corrosion Institute, 1983), pp. 48-52.

182. D.J. Greving, J.R. Shadley, "Experimental Evaluations of Thermal Spray Coatings for Oilfield Equipment Applications," in *Thermal Spray: International Advances in Coatings Technology, Proceedings of the International Thermal Spray Conference* (Materials Park, OH: ASM International, 1992), pp. 605-610.

183. E. Hayes, D.P. Thun, "Development of a Thermally Efficient Paper Machine Yankee Dryer Coating," in *Thermal Spray Coatings—New Materials, Processes and Applications, Proceedings of the 2nd National Conference on Thermal Spray* (Materials Park, OH: ASM International, 1984), pp. 21-26.

184. L. Suhl, National Thermal Spray Conference, Long Beach, CA (Materials Park, OH: ASM International, 1990).

185. R. Smith, R. Novak, *PMI* 23 3, 4(1991).

186. Coatings News, "Types of Hot Corrosion and Oxidation," *Materials Performance* 30, 6(1991): p. 47.

187. A. Kay, "Jet Kote Applications for Corrosion Protection," in *Thermal Spray—Advances in Coatings Technology, Proceedings of the National Thermal Spray Conference* (Materials Park, OH: ASM International, 1987), pp. 393-399.

188. P.E. Chandler, M.B. Quigley, "The Application of Plasma Sprayed Coatings for the Protection of Boiler Tubing," in *Advances in Thermal Spraying, Proceedings of the 11th International Thermal Spray Conference* (Montreal, Canada: Welding Institute of Canada, 1986), pp. 29-36.

189. T.C. Hurng et al., "The Development of a Hybrid Spraying System," in *Advances in Thermal Spraying, Proceedings of the 11th International Thermal Spraying Conference* (Montreal, Canada: Welding Institute of Canada, 1986), pp. 495-506.

190. T. Shinohara et al., U.K. Patent Application No. GB 2117415A.

191. M.L. Thorpe, R.H. Unger, "A New Coating for Corrosion Protection in Boilers," in *Advances in Thermal Spraying, Proceedings of the 11th International Thermal Spray Conference* (Montreal, Canada: Welding Institute of Canada, 1986), pp. 3-18.

192. S. Gustaffson, "Thermal Coating as Corrosion Protection in Boilers," in *Advances in Thermal Spraying, Proceedings of the 11th International Thermal Spray Conference* (Montreal, Canada: Welding Institute of Canada, 1986), pp. 19-28.

193. R.A. Miller, *TAPPI J.* (1984): p. 50.

194. J.H. Harrington, S. Rangaswamy, in *Coatings and Bimetallics for Aggressive Environments, Conference Proceedings* (Materials Park, OH: ASM International, 1985), p. 171.

195. W.C. Cochran, *Metal Progress* (1982): p. 37.

196. "Arc Sprayed Coatings to Combat High Temperature Corrosion" TAFA Internal Report 2.6.1.

197. Arcolin, Bertoni, Gaville, "Special Coatings for Ethylene Reactors Based on the Thermal Cracking of Hydrocarbons," 9th International Thermal Spray Conference (The Hague, Neth.: Nederlands Instituut voor Lastechniek, 1980), pp. 84-90.

198. K. Ishikawa et al., "Protection from High Temperature Corrosion with a Multi-Layer Thermal Spray Coating," in *Thermal Spray: International Advances in Coatings Technology, Proceedings of the International Thermal Spray Conference* (Materials Park, OH: ASM International, 1992), pp. 859-864.

199. S. Oki et al., "High Temperature Oxidation Properties of of Thermal Sprayed Ni-Cr-Al Multi-Layer Coatings," in *Thermal Spray Research and Applications, Proceedings of the 3rd National Thermal Spray Conference* (Materials Park, OH: ASM International, 1990), pp. 215-220.

200. K.F. Briegel, Rohm & Haas, correspondence to author, 1990.

201. S. Oki et al., "Diffusion Process between Thermal Sprayed Aluminum and Austenitic Stainless Steel, and Its Application," 12th International Conference on Thermal Spraying, paper no. 16 (London, UK: Welding Institute, 1989).

202. R.M. Kain, E.A. Baker, "Marine Atmospheric Corrosion Museum Report on the Performance of Thermal Spray Coatings on Steel," in *Testing of Metallic and Inorganic Coatings*, ASTM STP 947 (Philadelphia, PA: ASTM, 1987), pp. 211-234.

203. H. Herman, K.J. Altorfer, "Zinc Thermal Spray Metallizing for Corrosion Protection of Structural Steel," AMS Metals Congress, paper no. 8201-042 (Materials Park, OH: ASM International, 1982).

204. T. Gardega, ASM Regional Seminar on Thermal Spray Coating, SUNY at Stony Brook (Materials Park, OH: ASM International, 1986); see also T.W. Gardega et al., "Thermal Spray Coatings," *J. Protective Coatings & Linings* 3, 1 (1986).

205. Sulit, Kullerd, West, "Wire Sprayed Aluminum Coating Services in a Sima Corrosion Control Shop," in *Thermal Spray—Advances in Coatings Technology, Proceedings of the National Thermal Spray Conference* (Materials Park, OH: ASM International, 1987), pp. 241-252.

206. J.H. Payne, "Sealed Sprayed Metal Coatings for Economic Corrosion Protection of Structures," Symposium on Design with Thermal Spraying in the 80's, paper no. 5 (London, UK: Association of Metal Sprayers, 1982).

207. K. Elayaperunal, "Use of Coatings and a Special Stainless Steel in a Pulp and Paper Mill," in *Pulp and Paper Industry Corrosion Problems, Proceedings of the 4th International Symposium* (Stockholm, Sweden: Swedish Corrosion Institute, 1983), pp. 271-274.

208. J.C. Bailey, F.C. Porter, M. Round, "Metal Spraying of Zinc and Aluminum in the United Kingdom," 12th International Conference on Thermal Spraying, paper no. 8 (London, UK: Welding Institute, 1989).

209. Round, "The Development and Use of Electric Arc Sprayed Aluminum Coatings for Corrosion Protection," 9th International Thermal Spray Conference (The Hague, Neth.: Nederlands Instituut voor Lastechniek, 1980), pp. 392-399.

210. C.P. Dillon, *Materials Performance* 29, 11 (1990): pp. 54-55.

211. Basaraba, Bakst, *J. Protective Coatings and Linings* (Nov. 1990): p. 8.

212. A. Ohliger, P. Szelagowski, "Thermal Spraying—Recent Developments for Underwater Application," in *Thermal Spray Research and Applications, Proceedings of the 3rd National Thermal Spray Conference* (Materials Park, OH: ASM International, 1990), pp. 193-200.

213. H. Morrow, Zinc Metallizers Task Group, correspondence to author, 1989

214. M.E. Leclerq, R. Bensimone, "New Zinc-Based Alloy for Metallizing," 8th International Thermal Spraying Conference (Essen, FRG: German Welding Society, 1976), pp. 417-429.

215. Z. Zhongli, G. Weisheng, W. Jinlin, "The Arc Sprayed Zn-Al Pseudo-Alloy Coating," in *Thermal Spray: International Advances in Coatings Technology, Proceedings of the International Thermal Spray Conference* (Materials Park, OH: ASM International, 1992), pp. 399-403.

216. Duralcan product literature, 1991.

217. R.P. Krepski, N.G. Feige, "Zinc-Nickel Thermal Spray Coating for Mitigation of Stress Corrosion Cracking and Corrosion of Steel in Concrete," International Thermal Spray Conference (Materials Park, OH: ASM International, 1992).

218. C. Reynolds, National Coatings, correspondence to author, 1992.

219. H. Morrow, 1992 International Thermal Spray Conference (Materials Park, OH: ASM International, 1992).

220. R.L. Jones, "High Temperature Vanadate Corrosion of Yttria-Stabilized Zirconia Coatings on Mild Steel," in *Thermal Spray Technology—New Ideas and Processes, Proceedings of the National Thermal Spray Conference* (Materials Park, OH: ASM International, 1988), pp. 283-288.

221. C. Berndt et al., "Fire Barrier Coatings for Protection of Aluminum Structures," 10th International Thermal Spray Conference (Essen, FRG: German Welding Society, 1983).

222. K. Asano et al., "Latest Progress and Future Development of Flame Gunning Technique," Nippon Steel Company technical report No. 41 (April 1989), pp. 52-58.

223. W.J. Lenling, M.F. Smith, D.I. Orloff, "Thermal Coating Development for Impulse Drying," in *Thermal Spray: International Advances in Coatings Technology, Proceedings of the Inter-*

national Thermal Spray Conference (Materials Park, OH: ASM International, 1992), pp. 619-624.

224. C. Berndt, "Mechanical, Thermomechanical, and Tribological Applications of Ceramic Coatings," 7th CIMTEC World Ceramic Congress, Montecatani, Italy.

225. H. Noji et al., "A Study of Thermal Spraying of Glass and Its Mechanical Properties," 12th International Conference on Thermal Spraying, paper no. 14 (London, UK: Welding Institute, 1989).

226. R. Polak, L.G. Witts, "Roughness Adapted Surface Layers," 12th International Conference on Thermal Spraying, paper no. 87 (London, UK: Welding Institute, 1989).

227. T.N. Rhys-Jones, "Applications of Thermally Sprayed Coating Systems into Aero Engines," 12th International Conference on Thermal Spraying, paper no. 90 (London, UK: The Welding Institute, 1989).

228. R.A. Perkins, 1990 National Thermal Spray Conference, Long Beach, CA.

229. Copperlok Inc. product literature.

230. L.M. Wyatt et al., *An Atlas of Corrosion and Related Microstructures*, MTI Publication No. 18 (St. Louis, MO: Materials Technology Institute, 1987). [This book is available from NACE International.]

231. Process Industries Corrosion (Houston, TX: NACE International, 1975).

232. ASM *Handbook*, 9th ed., vol. 13, "Corrosion," (Materials Park, OH: ASM International, 1987).

233. ASM Handbook, vol. 18, *Friction*, "Lubrication and Wear Technology" (Materials Park, OH: ASM International, 1992).

234. H. Nakahiri et al., "Advanced Thermal Barrier Coatings Involving Efficient Vertical Micro-Cracks," in *Thermal Spray: International Advances in Coatings Technology, Proceedings of the International Thermal Spray Conference* (Materials Park, OH: ASM International, 1992), pp. 519-524.

BIBLIOGRAPHY

Adamski, A., R. McPherson, "Cavitation Erosion Behavior of Several Thermally Sprayed Coatings," in *Advances in Thermal Spraying, Proceedings of the 11th International Thermal Spraying Conference*. Materials Park, OH: ASM International, 1986.

Arata, Y., A. Ohmori, C.J. Li, "Study on the Structure of Plasma Sprayed Ceramic Coating Using Copper Electroplating," in *Proceedings of the International Symposium on Advanced Thermal Spraying Technology and Allied Coatings*. Osaka, Japan: High Temperature Society of Japan, 1988.

Berndt, C., ed. *Thermal Spray: International Advances in Coatings Technology*. Materials Park, OH: ASM International, 1992.

Berndt, C., T. Bernecki, eds. *Thermal Spray Coatings: Research, Design and Appl.ications*. Materials Park, OH: ASM International, 1993.

Bernecki, T., ed. *Thermal Spray Coatings: Properties, Processes & Applications*. Materials Park, OH: ASM International, 1991.

Drossman, R.O., R.P. Krepski, "Tribolite Iron-Based Thermal Spray Coatings for the Control of Wear," in *Thermal Spray Coatings: Properties, Processes and Applications, Proceedings of the 4th National Thermal Spray Conference*. Materials Park, OH: ASM International, 1991.

Houben, J.M., "Future Developments in Thermal Spraying," in *Thermal Spray Coatings—New Materials, Processes and Applications, Proceedings of the 2nd National Conference on Thermal Spray*. Materials Park, OH: ASM International, 1984.

Houk, Cheney, Cribbs, "Properties of Oxygen-Bearing Molybdenum Coatings," Proceedings of the 9th International Thermal Spray Conference. The Hague, Neth.: Nederlands Instituut voor Lastechniek, 1980.

Journal of Thermal Spray Technology, Materials Park, OH: ASM International.

Marantz, D.R., "Electromagnetically Coalesced Multi-Arc Plasma Torch with True Axial Powder Feed," in *Thermal Spray Research and Applications, Proceedings of the 3rd National Thermal Spray Conference*. Materials Park, OH: ASM International, 1990.

McPherson, R., Cheang, P., "Microstructural Analysis of Ni-Al Plasma Sprayed Coatings," in *Proceedings of the 12th International Conference on Thermal Spraying*, paper no. 17. London, UK: The Welding Institute, 1989.

Meringolo, V., ed. *Thermal Spray Coatings*. Atlanta, GA: TAPPI Press, 1983.

Moscowitz, L.N., "Application of HVOF Thermal Spraying to Solve Corrosion Problems in the Petroleum Industry," in *Thermal Spray: International Advances in Coatings Technology, Proceedings of the International Thermal Spray Conference*. Materials Park, OH: ASM International, 1992.

Rao, K.V., Somerville, D.A., Lee, D.A., "Properties and Characterization of Coatings Made Using Jet Kote Thermal Spray Technique," in *Advances in Thermal Spraying, Proceedings of the 11th International Thermal Spraying Conference*. Montreal, Canada: Welding Institute of Canada, 1986.

Sampath, S., Herman, H., Rangaswamy, S., "Ni-Al Re-Evaluated," in *Thermal Spray—Advances in Coatings Technology, Proceedings of the National Thermal Spray Conference*. Materials Park, OH: ASM International, 1987.

Shaw, B.A., Morton, A.G., "Thermal Spray Coatings-Marine Performance and Mechanisms," in *Thermal Spray Technology—New Ideas and Processes, Proceedings of the National Thermal Spray Conference*. Materials Park, OH: ASM International, 1988.

Thermal Spraying Practice, Theory, and Applications. Miami, FL: American Welding Society, 1985.

Tucker, R.C., "Plasma and Detonation Gun Deposition Techniques and Coating Properties," in *Deposition Technologies for Films and Coatings*. Park Ridge, NJ, Noyes Publications, 1982.

Unger, R., "Comparison of Thermal Spray Bond Coats," in *Thermal Spray—Advances in Coatings Technology, Proceedings of the National Thermal Spray Conference*. Materials Park, OH: ASM International, 1987.

Wilms, V., Herman, H., "Crystallography and Microstructure of Equilibrium and Metastable Phases Resulting from Plasma Spraying of Ceramic Coatings," in *Proceedings of the 8th International Thermal Spraying Conference*. Miami Beach, FL: American Welding Society, 1976.

INDEX